# Animals of the Oceans
## the ecology of marine life

# Animals of the Oceans
## the ecology of marine life

by Martin Angel & Tegwyn Harris

Peter Lowe

Previous page:
Seahorses cling to weeds or corals with their tails and swim vertically, sculling with their dorsal fins.

A storm approaching from the sea drives waves ashore. Salts lost in spray and geochemical processes in the sediments of the ocean floor are replaced by the constant inflow of rivers and weathering of rocks on land.

Index compiled by M. O'Hanlon



ISBN 0 85654 612 7

Filmset by Typesetting Services Ltd.
Glasgow and Edinburgh
Printed in Great Britain by
William Clowes & Sons, Limited
London, Beccles and Colchester

# Introduction

Almost 71 % of the earth's surface is covered by the sea. In terms of actual measurement, this is approximately 361,000,000 km² in area. In places the sea is over 10,000 m deep—the deepest point so far reliably measured being 11,033 m.

This vast volume of water, a complex solution of many chemical substances, is constantly subjected to a wide range of physical and meteorological forces. Due to the rotation of the earth and the gravitational influence of the sun and the moon, tides and waves are set in motion and currents are generated which agitate the water from the deepest abyss to the bordering seashore. Although the deepest parts of the ocean are constantly dark and cold, the shallower layers are subjected to appreciable heating in equatorial regions; at the poles, the ambient temperature is so low that the aqueous part of the sea freezes. Even the salinity of the sea is not constant but varies from place to place and often from season to season.

Such a multiplicity of environmental conditions gives rise to a corresponding multitude of ecological habitats which support a bewildering variety of plants and animals, ranging in size from microscopic algae to gigantic fishes and whales.

As in all natural ecosystems, the success of the animal population as a whole depends, ultimately, upon primary production of photosynthesized food. Whilst the large sea-weeds are undoubtedly the most familiar marine plants, they contribute only a relatively tiny amount to the required food store. All primary production significant to the ocean economy takes place in the phytoplankton—the great meadow of microscopic and sub-microscopic plants which live in the illuminated surface layers. Here, too, lives the greater part of the zooplankton—microscopic animals, many of which are secondary herbivorous producers and form the first step in an extensive, complex food-web.

All the ocean's different animals, whether they are herbivores or primary, secondary or tertiary carnivores depend upon the phytoplanktonic meadows. Many of these oceanic populations are of economic importance to man. An increasing awareness of their ecological requirements and limitations has, over the years, led to an increased use of scientific methods to study their biology and the influence of the physical properties of the oceans upon them. The information thus gathered can be used both to protect and to exploit them.

As the open sea is of such enormous area it has a correspondingly long boundary. The seashore, as understood by marine biologists, is that area of land between and just beyond the highest and the lowest limits of extreme spring tides. The total area of seashore defies imagination, as it virtually defies mathematical measurement, but an idea of its scale may be obtained from an estimate of the shore area of the British Isles at 2,510 km².

Since this ribbon of land extends through cold, temperate and hot latitudes; is subjected to regular inundation by the sea and to the rigours of aeriel desiccation and heat; is sometimes pounded by almost unimaginable forces and is in places washed by fresh water, it is not surprising that it, too, like the open sea is made up of an enormous range of habitats. The adaptations which have led to the colonization of these by plants and animals are often strange, always interesting and sometimes bizarre.

Amongst the most interesting and complex of the littoral habitats are those which in the tropics support mangrove swamps and coral reefs. These are especially important since they in turn provide further ecological niches which lead, inevitably, to very complex associations.

Most bizarre of all are undoubtedly the adaptations of the inhabitants of the other oceanic 'boundary'—the deep sea and its floor. Here, where environmental conditions are quite unlike any others on earth, the extreme habitats are populated by animals which themselves show extremes of evolutionary development.

Since the sea covers so much of the earth's surface it is inevitable that resources of food and deposits of industrial raw materials lie within it. Despite the obvious problems involved, man has turned his attention increasingly to finding ways of tapping these resources and of using the sea for a number of purposes. Though useful to modern civilization, this may not always be to the sea's (and its animals') advantage.

This book seeks to survey some of these ecological matters and to present a picture of the sea as a complex of interrelated but diverse habitats occupied by a wealth of animal life.

*Martin Angel*
*Tegwyn Harris*

# The Ocean Environment

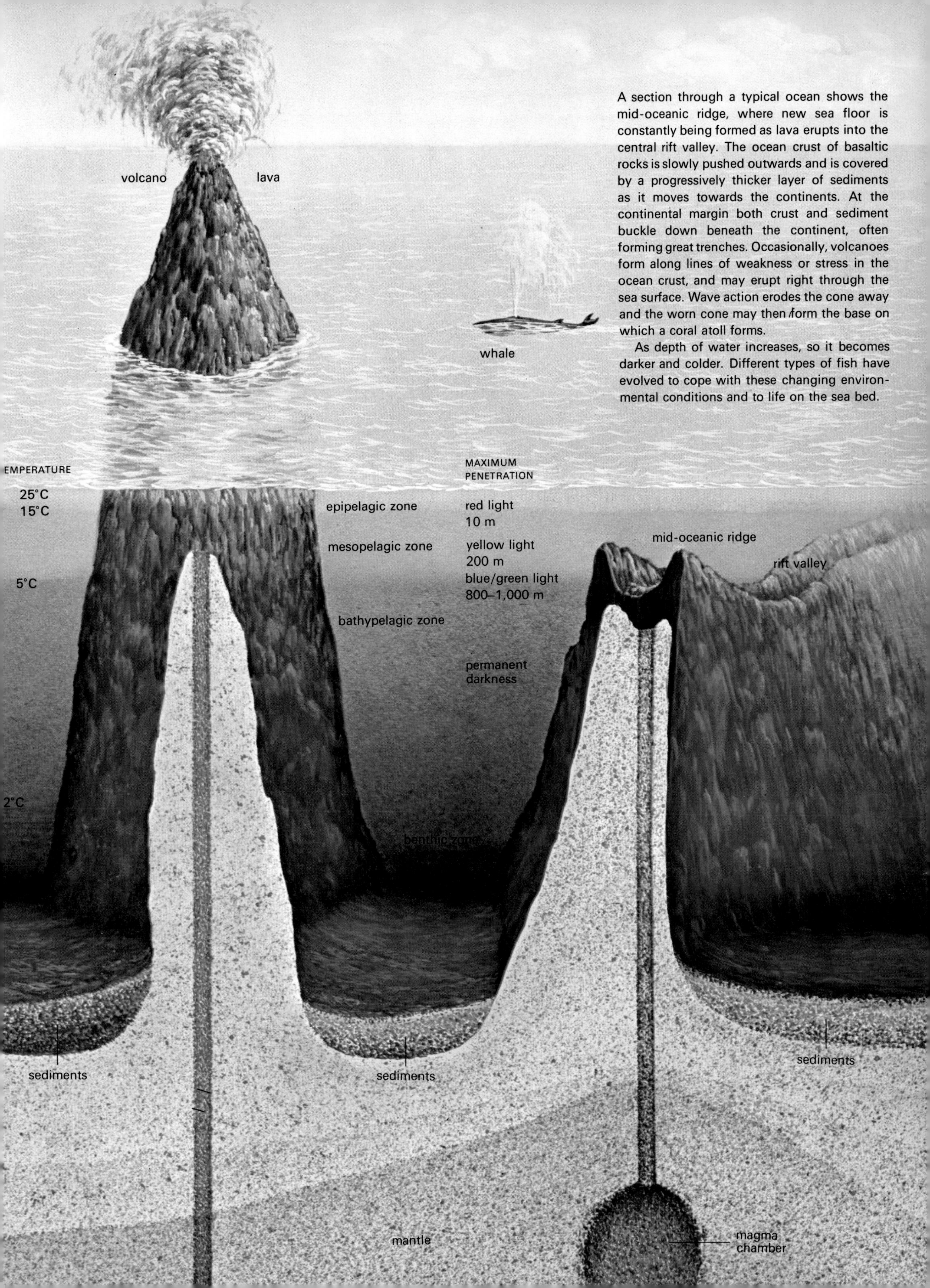


A section through a typical ocean shows the mid-oceanic ridge, where new sea floor is constantly being formed as lava erupts into the central rift valley. The ocean crust of basaltic rocks is slowly pushed outwards and is covered by a progressively thicker layer of sediments as it moves towards the continents. At the continental margin both crust and sediment buckle down beneath the continent, often forming great trenches. Occasionally, volcanoes form along lines of weakness or stress in the ocean crust, and may erupt right through the sea surface. Wave action erodes the cone away and the worn cone may then form the base on which a coral atoll forms.

As depth of water increases, so it becomes darker and colder. Different types of fish have evolved to cope with these changing environmental conditions and to life on the sea bed.

A casual glance through an atlas gives the impression that the Earth's surface is dominated by land, yet a thorough examination shows that 70·8 % is covered by ocean. The land masses divide this great area of the seas into three major oceans, the Pacific, Atlantic and Indian. The Pacific Ocean, with an area of 165·25 million square kilometres and the Atlantic Ocean, with an area of 82·22 million square kilometres, both open into the smaller Arctic and Antarctic Oceans. Most of the 73·44 million square kilometres of the Indian Ocean lie in the southern hemisphere and connect only with the Antarctic. Marginal seas, like the North Sea and the 'Mediterranean' type seas, which include the Caribbean, the Baltic and the Red Sea, provide a further 40 million square kilometres. In comparison, the approximate areas of the great continents in millions of square kilometres are: Eurasia 52·20, the Americas 41·86, Australasia 9·94 and Africa 29·75.

Not only are the oceans very extensive but they are deep: 53·6 % of the Earth's surface is covered with water to a depth of 3,000–6,000 m. The mean depth is 3,790 m, giving a total estimated volume for the whole ocean of 1,367 million cubic kilometres. However, this almost inconceivably huge volume provides only 0·24 % of the Earth's mass. It is unquestionably the largest environment available for life and, apart from a few exceptional areas, is inhabited throughout its volume.

The Earth is unique amongst the planets of the solar system in possessing so much water. It is large enough for its gravitational field to hold an atmosphere and because of its distance from the sun, the surface temperature is cool enough for the water vapour to condense into liquid. It is not, however, cold enough for the water to freeze into ice over more than a small part of the Earth's surface.

The volume of the oceans has remained remarkably constant throughout the geological past. This consistency is a mystery since it might be expected that the water would gradually become separated into oxygen and hydrogen and be lost from the Earth's atmosphere as neon has been.

## The origin of the oceans

Despite the constancy of the volume of the oceans, their shape, depth and distribution have changed throughout geological history. The geology of the ocean floor is fundamentally different from the continental land masses, where rocks as old as 2,000 million years are not uncommon. The oldest rocks so far discovered occur in Greenland and radioactive dating gives them an age of 3,800 million years. The ocean bed is formed of crustal rocks overlying the Earth's mantle. The lower levels of the crustal rocks consist of gabbro, which is overlain by basaltic dykes, vertically packed sheets of rock, then pillow lavas, and finally oceanic sediments. The most ancient of these are little more than 150 million years old. The explanation for this difference in ages was accounted for by a simplistic theory put forward by a palaeontologist named Wegener. Wegener noticed that the shapes of many of the continents could be fitted together well. Furthermore, the continents that did fit together well, such as South America and Africa, had similar ancient fossil faunas. He suggested that the continents were once joined into one super-continent called Pangaea and had broken loose and drifted apart; new ocean floor had formed at the same time. Reaction to this theory was sceptical in the extreme for half a century, the common fossil faunas being explained by land bridges and lost continents like Atlantis. However, the theory is now accepted in a modified form as the theory of plate tectonics which divides the surface of the earth into a number of rigid, moving plates on which the continents are carried.

## Plate tectonics and sea-floor spreading

Typically, down the centre of each major ocean there runs a chain of mountains called the mid-oceanic ridge. Along the centre of each mid-oceanic ridge there is a split in the ocean floor formed by a great rift valley. These rift valleys are usually about 20 km wide and 1 km deep. The sides are very steep, falling in two or three massive steps with slopes of 70–80° formed by series of faults. Along the centre of the bed of a typical valley, observers in special deep sea submersibles have seen and photographed great piles of solidified pillow lava. Lava which erupts underwater is cooled very rapidly and solidifies into globular masses a metre or so across. Sound waves generated by underwater explosions are used to penetrate and explore the structure of the rocks beneath the rift valleys. Present results suggest that the lavas have been forced up through dykes or volcanic tubes from an underlying magma chamber, solidifying as they cool into great pillow-like lumps.

Borings taken into the sea-bed have shown that

HEATHER ANGEL

The friction of wind blowing over the sea whips up the surface into waves. These help to churn up the near surface layers of the sea, erode coastlines and affect the survival of coastal species.

The map on the following pages shows the major physical features of the oceans.

the further away from the mid-oceanic ridge the cores are taken, the thicker the layer of sediment that overlies the basaltic rock. Analysis of fossil remains of planktonic animals and plants found in these sediments reveals several interesting features. Firstly, the thicker the sediments, the older are the fossils in the lower layers. Secondly, accumulation rates in most oceanic areas are less than a millimetre per thousand years, although very close to the continents they may be as much as a centimetre per century. The impression is gained that the oldest ocean floor occurs along the edges of the seas and the youngest in the region of the mid-oceanic ridge. This has been confirmed in the Central Atlantic by boring right through the overlying sediments to the basalts beneath and radioactively dating these basal rocks.

The central rift valleys are centres for much earthquake activity and it now seems likely that these earthquakes are associated with the constant formation of new sea-floor. The new sea-floor pushes, or is pushed, out sideways from the mid-oceanic ridge as a rigid plate which may be 70–100 km thick. At the boundary of ocean and continent, where two plates meet, one of two phenomena occur. Either the continental land

In the open ocean the bottom muds are often composed purely of the skeletons of planktonic organisms. Here, greatly enlarged, is a sample of radiolarian ooze, formed from the accumulated glassy silica skeletons of single-celled protozoan animals.

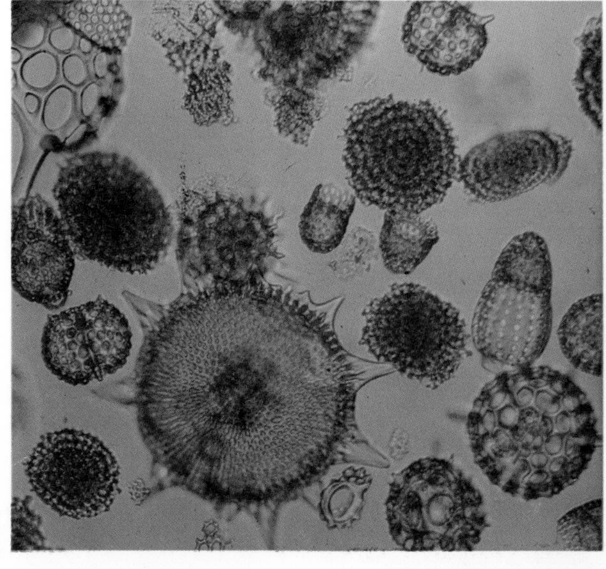

CLEGG

160°
120°
80°
40°
0°
80°
40°
0° EQUATOR
40°
Laurentian
ARCTIC OCEAN
Basin
LINCOLN SEA
WANDELS SEA
NANSEN RISE
ELLESMERE ISLAND
BEAUFORT SEA
Land's End
Beaufort Basin
BANKS I.
VISCOUNT MELVILLE SD.
VICTORIA I.
AMUNDSEN G.
G. OF BOOTHIA
BAFFIN ISLAND
BAFFIN
Baffin Basin
BAY
DAVIS STR.
GREENLAND
GREENLAND SEA
SCORESBY SD.
Norweg
NORWEGI.
SEA
Basin
CHUKCHI SEA
BERING STR.
ARCTIC CIRCLE
NORTON SD.
FOXE
BASIN
DENMARK STR.
ICELAND
HUDSON BAY
C. Chidley
C. Farewell
REYKJANES RIDGE
4316
GULF OF ALASKA
BRISTOL BAY
7250
Aleutian Trench
415
NORTH
BRITISH ISLES
PORCUPINE BANK
West European
Basin
Land's End
ENGLISH CH.
NEWFOUNDLAND
C. Flattery
GULF OF ST. LAWRENCE
C. Race
GRAND NEWFOUNDLAND BANKS
NEWFOUNDLAND RISE
NORTH
6325
BAY OF BISCAY
759
MENDOCINO SEASCARP
2806
C. Mendocino
C. Cod
MUIR SEAMOUNT 1417
Basin
Iberian Basin
C. St. Vincent
AZORES-C. ST VINCENT RIDGE
STRAIT OF GIBRALTAR
AMERICA
C. Hatteras
North-Western Atlantic
772
ERBEN TABLEMOUNT 412
MURRAY SEASCARP
ATLANTIC
6474 Murray Deep
C. Kennedy
6905 Nares Deep
Canaries Basin
CANARY IS.
TROPIC OF CANCER
G. OF CALIFORNIA
GULF OF MEXICO
HAWAIIAN RIDGE
HAWAIIAN IS.
WEST
Milwaukee Depth 9220
INDIES
GREATER ANTILLES
OCEAN
MID-ATLANTIC RIDGE
6216
Clarion Fracture Zone
Acapulco Tr.
CARIBBEAN SEA
Venezuelan Basin
Cape Verde
Cape Verde Basin
6662
Guatemala Tr.
Guatemala Basin
6182
Clipperton Fracture Zone
115
N.W. CHRISTMAS ISLAND RIDGE
20
ALBATROSS PLATEAU
COCOS RIDGE
G. OF PANAMA
2276
SOUTH
Sierra Leone Basin
GULF OF GUI
Guinea Basin
GALÁPAGOS IS.
7310
Romanche Gap
PACIFIC
Aguja Pt.
Cape São Roque
EAST PACIFIC RIDGE
Peru-Chile Trench
AMERICA
TUAMOTU RIDGE
1929
Peru Basin
Brazilian Basin
SOUTH
MID-ATLANTIC RIDGE
South-Eastern Atlantic Basin
SOUTH-EASTERN PACIFIC PLATEAU
TROPIC OF CAPRICORN
AUSTRAL RIDGE
SAN FELIX JUAN FERNANDEZ RIDGE
Richards Deep 7635
Cape Frio
ATLANTI
WALVIS RI
OCEAN
730
South-Western
C. Santa María
1528
Argentine Basin
OCEAN
Pacific Basin
PACIFIC-ANTARCTIC RIDGE
Pacific-Antarctic Basin
ATLANTIC-ANTARCTIC
ATLANTIC-INDIAN RIDGE
STR. OF MAGELLAN
FALKLAND IS.
SCOTIA RIDGE
SOUTH GEORGIA
8264
Meteor Depth
South Sandwich Trench
168
Scotia Basin
SCOTIA SEA
C. Horn
5832
5140
675
DRAKE PASSAGE
SCOTIA RIDGE
1874
Pacific
Atlanti
SOUTHERN OCEAN
4951
ANTARCTIC CIRCLE
Equatorial Scale
Miles
0
1000
2000
3000
0
1000
2000
3000
4000
Kilometres
AMUNDSEN SEA
WEDDELL SEA
© Geographical Projects

40°
80°
120°
160°
ARCTIC OCEAN
FRANZ JOSEF LAND
SEVERNAYA ZEMLYA
SPITSBERGEN
C. Chelyuskin
NEW SIBERIAN ISLANDS
EAST SIBERIAN SEA
LAPTEV SEA
NOVAYA ZEMLYA
BARENTS SEA
KARA SEA
North Cape
CHUKCHI SEA
ARCTIC CIRCLE
G. OF BOTHNIA
BALTIC SEA
C. Navarin
SEA OF OKHOTSK
BERING SEA
C. Yelizaveta
ALEUTIAN IS
EURASIA
SAKHALIN
C. Lopatka
Aleutian Trench
Kurile Trench
Vityaz Deep
10,377
BLACK SEA
CASPIAN SEA
SEA OF JAPAN
HOKKAIDO
HONSHU
EMPEROR SEAMOUNTS
YELLOW SEA
Japan Trench
PACIFIC
RANEAN SEA
Ramapo Deep
10,500
EAST CHINA SEA
Nansei Shoto Tr.
SOUTH HONSHU RIDGE
HAWAIIAN RIDGE
PERSIAN G.
TROPIC OF CANCER
TAIWAN
KYUSHU PALAU RIDGE
Mariana Trench
RED SEA
ARABIAN SEA
SOUTH CHINA SEA
MAGELLAN SEAMOUNTS
MARCUS-NECKER RISE
PHILIPPINE ISLANDS
ICA
BAY OF BENGAL
BENGAL PLATEAU
GULF OF ADEN
SOCOTRA
Arabian Basin
ANDAMAN SEA
GULF OF SIAM
Challenger Depth
10,863
C. Guardafui
BENGAL
Cape Johnson Depth
10,500
Philippine Trench
C. Comorin
SRI LANKA
MALDIVE RIDGE
CARLSBERG RIDGE
West Caroline Basin
East Caroline Basin
CAROLINE SOLOMON RIDGE
OCEAN
Somali Basin
CELEBES SEA
NINETYEAST RIDGE
Mentawai Trench
SUMATRA
MENTAWAI RIDGE
BORNEO
CELEBES
EQUATOR
0°
MASCARENE RIDGE
Nazareth Trough
JAVA SEA
BANDA SEA
NEW GUINEA
SOLOMON ISLANDS
JAVA
Mid-Indian Basin
JAVA RIDGE
Java Trough
TIMOR
ARAFURA SEA
York
Mascarene Basin
C. Ambre
TIMOR SEA
GULF OF CARPENTARIA
MELANESIAN BORDER PLAT.
MOZAMBIQUE CHAN.
INDIAN OCEAN
MID-INDIAN RIDGE
Wharton Deep
6459
West Australian Basin
CORAL SEA
North Fiji Basin
MADAGASCAR
N. West C.
W. AUSTRALIAN RIDGE
BELLONA PLATEAU
HUNTER RIDGE
Tonga Tr.
TROPIC OF CAPRICORN
AUSTRALIA
10,633
Horizon Depth
SOUTH-EAST INDIAN RIDGE
NORFOLK I. RIDGE
South Fiji Basin
KERMADEC RIDGE
Kermadec Tr.
S. MADAGASCAR RIDGE
NATAL RIDGE
Natal Basin
5788
C. Leeuwin
GREAT AUSTRALIAN BIGHT
NATURALISTE RIDGE
Ulladulla Trough
Galathea Depth
9994
SOUTH-WEST INDIAN RIDGE
South-East Indian Basin
South Australian Basin
LORD HOWE RISE
BASS STR.
TASMANIA
NEW ZEALAND
TASMAN SEA
CHATHAM RISE
AMSTERDAM-ST. PAUL PLATEAU
Agulhas Basin
PR. EDWARD-CROZET RIDGE
C. Providence
S. TASMANIA RIDGE
KERGUELEN IS.
INDIAN-ANTARCTIC RIDGE
KERGUELEN-GAUSSBERG RIDGE
Indian-Antarctic Basin
BANZARE SEAMOUNT
188
5872
Eastern Indian-Antarctic Basin
309
GRIBBS SEAMOUNT
MACQUARIE-BALLENY RIDGE
5496
SOUTHERN OCEAN
ANTARCTIC CIRCLE
ANTARCTICA
ROSS SEA

mass rides with the moving plate and so slowly drifts across the Earth's surface, or the plate slides beneath the continent. Along the Pacific coast of South America, the plate dives down at an angle of 45°, causing frequent and violent earthquakes, such as the one that destroyed the city of Managua, the capital of Nicaragua, in 1973.

Often great trenches are formed as the ocean floor buckles down under the continental land mass or another tectonic plate. The deepest sounding (11,515 m) in the world's oceans was recorded in the Mindanao Trench, just off the Philippines. There is much volcanic activity in such areas and active mountain-building occurs. The Andes are still being buckled up by the Pacific plate pushing beneath and are volcanically highly active. The Ural mountains, in the centre of Russia, are similar in their geomorphology to the Andes and this suggests that the Urals were once on the border of an ocean that divided what is now Siberia from western Russia. Indeed the remains of an oceanic plate swallowed up 250 million years ago has been identified beneath the Urals from seismic records. Similarly the Appalachians in North America consist of sedimentary rocks laid down on the bed of an ancient ocean 400–650 million years ago. A hundred and fifty million years ago the Atlantic did not exist.

The Mediterranean area is influenced by the movement of the African plate relative to the Eurasian plate. Borings taken in the Mediterranean show that there are extensive salt deposits beneath the surface sediments and it now seems likely that, in the past, the Mediterranean has completely dried out at least once, if not twice. This was probably due to the Straits of Gibraltar being closed by the movements of the African plate and all the water evaporating; a process which, under present weather conditions, would only take a few thousand years. The Mediterranean would then have refilled in a catastrophic flood through the Straits of Gibraltar. Such an inflow of water would have suddenly dropped world-wide sea levels many tens of metres, probably wiping out all shore faunas.

The rates at which sea-floor spreading occurred in geological time can be estimated from the residual magnetism of the underlying rocks. As the lavas solidify when they erupt in the mid-oceanic rift valley, the rock crystals are magnetized by the Earth's magnetic field. If a research ship measures the strength of the Earth's magnetic field, the measurement is therefore greater than expected because the Earth's magnetic field is supplemented by the magnetism of the rock crystals. This is termed a positive anomaly. At irregular intervals of 50,000 to a million years or more, the polarity of the Earth's magnetism reverses; a compass needle would then point to the south instead of the north. The magnetism of the lavas formed before the reversal would then antagonize the Earth's field, reducing its observed strength and producing a negative anomaly. These alternating bands of positive and negative anomalies give a rough time scale of the rate at which the sea-floor has spread, in the form of a sort of magnetic tape recording. The bands run parallel with the mid-oceanic ridge but become distorted by faulting which is brought about by the Earth's solid geometry and by the gradual rotation of some of the plates.

Using these bands of magnetic anomalies it can be shown that the maximum rate of sea-floor spreading is 16 cm per year. The subcontinent of India was once joined to Africa. It broke off and 'drifted' eastwards, opening up what is now the Western Indian Ocean at a rate of 7 cm per year, and leaving behind it a piece which is Madagascar. India's collision with the Asian land mass about 50–70 million years ago caused the upfolding of the Himalayan mountain range. Earthquakes in the Himalayan region indicate that this process is continuing and that India still has not stopped moving. Sometime during this titanic movement, a huge fault or fracture zone was created. The Carlsberg Ridge is the mid-oceanic ridge of the Gulf of Aden, which is an ocean in embryo.

## Sea-level changes

Volcanoes, active and extinct, are common features of the ocean floor. They thrust up through the sediments, often right up to and through the sea surface, to form islands. The volcanic cone is then cut back and eroded away by wave action to form flat topped underwater sea-mounts or *guyots*. The flat tops of the guyots represent ancient sea-levels.

The sea-level changes not only through dramatic events, like the drying up and reflooding of the Mediterranean, but also through the slow accumulation of deep sea sediments and the effects of the ice ages. Even today, if the great ice-caps of Greenland and the 3,500 m thickness of ice on central Antarctica melted, the world sea-level would rise at least 50 m. The great weight of ice on Antarctica has depressed the level of the continent so that its shallow shelf seas are over 300 m deep

HAWKES/NHPA

Iceland lies on the mid-oceanic ridge of the North Atlantic, and so is volcanically highly active. In 1963 a new island, Surtsey, was formed by a massive eruption, a process by which many oceanic islands were created in the zoological past.

HUTSON/NSP

Sea levels vary either tidally or climatically through increase or melting of ice caps or geologically by earth movements. Coral islands are formed by the lowering of sea levels and the uplift of reefs. Wave action then erodes away the coralline rock. Round Island, Aldabra, in the Indian Ocean, is now the site for a nesting colony of noddy terns. Below: At Urvina Bay, Galapagos, 2 $km^2$ of sea floor were uplifted in 1954, stranding all the encrusting barnacles and tube worms.

HEATHER ANGEL

compared with 200 m of other continents. If the ice melted, the Earth's crust would slowly ease back to near its original position. This type of movement is taking place in western Europe.

## Atolls and sea-mounts

Similar earth movements are important in the formation of coral atolls. An underwater volcano erupts and forms an island. The waves start to erode the cone and the weight of the volcano depresses the Earth's crust, so that the volcano sinks back below the waves. However, the eroding rocks form a platform on which corals begin to grow. The increasing weight of coral adds to the weight of the volcano so that the crust is depressed even more. More coral grows up on top, and so the process continues. Borings taken into Eniwetok Atoll, one of the Marshall Islands in the Pacific, showed a thickness of over 1,500 m of coralline rock, the result of millions of years of coral growth.

Not all sea-mounts are volcanoes. Rockall Bank in the North Atlantic is composed of sedimentary rock and is thus a piece of continent that broke away from the European land mass as the Atlantic opened up. Similarly the Seychelles Islands in the Western Indian Ocean are granitic and of continental origin; the deep area between the Seychelles and the African mainland, since it is totally lacking in magnetic anomalies, has the properties of a sunken continental area rather than of a typical ocean floor.

## Sea-water

Water is the main constituent of the sea. We have no geological record of its origin, but it was probably released as a liquid from chemical compounds during the initial thousand million years of the Earth's history. It is a remarkable chemical compound. Freshwater, when it is cooled, becomes progressively denser until it reaches a temperature of 4°C, below which it becomes lighter until it freezes. However, when salt is dissolved in it, the freezing point is lowered and the density continues to increase until the freezing point is reached. Sea-water freezes at −1·9°C.

In ice, the water molecules associate into a structure which increases the volume and hence decreases the density of the ice relative to unfrozen water by 9%. Thus ice floats on water and so acts as an insulator, slowing down the further cooling and eventual freezing of the water lying below.

## Salts in sea-water

Another important chemical property of water is that it is an excellent solvent, especially for inorganic salts. The quality of the salts dissolved in sea-water is very consistent. Derived from the constant weathering and leaching of rocks on the continental land masses, the various constituents now appear to be in a state of equilibrium: new salts are constantly being added through the inflow of rivers while others are removed in spray and by geochemical processes within the sediments. Sea-water is not just a solution of common salt (sodium chloride) but a solution of a complex mixture of salts. Since marine animals utilize many of the minor constituents, even if a marine aquarium is filled with a common salt solution of the right concentration, all its inhabitants will soon die.

Salts, when dissolved in water, separate into their main constituents (ions). There are two types of ions; cations are positively charged and act as alkalis, whereas the negatively charged anions act as acids. The major cations in sea-water are sodium, magnesium, calcium, potassium, and strontium. The major anions are: chloride, sulphate, bicarbonate, and bromide.

Oceanic animals have a ready supply of both calcium and bicarbonate which are easily transformed into insoluble calcium carbonate. This is widely used by both animals and plants for skeletal structures. Animals using it include the microscopic Foraminifera, molluscs, echinoderms (e.g. sea-urchins) and fish; the plants include the coccolithophores whose minute fossilized remains form chalk. Down to depths of about 4,000 m these calcareous skeletons are a major constituent of the sediments that accumulate so slowly. These are some of the microfossils that are used by palaeontologists to trace ancient planktonic faunas. However, below about 4,000 m, the hydrostatic pressure produced by the overlying weight of water begins to reverse the chemical equilibria that keep calcium carbonate insoluble and it begins to dissolve. Below this depth, which is known as the *carbonate compensation depth*, the bottom deposits are red clays which contain no carbonate skeletons.

Sea-water also contains minute traces of numerous other elements, such as iron, copper, vanadium, manganese, cobalt, nickel and even gold. Often the quantities are exceedingly small; gold, for example, is present in concentrations of 0·004 mg per cubic metre of sea-water. Even so, some of these elements are extremely important

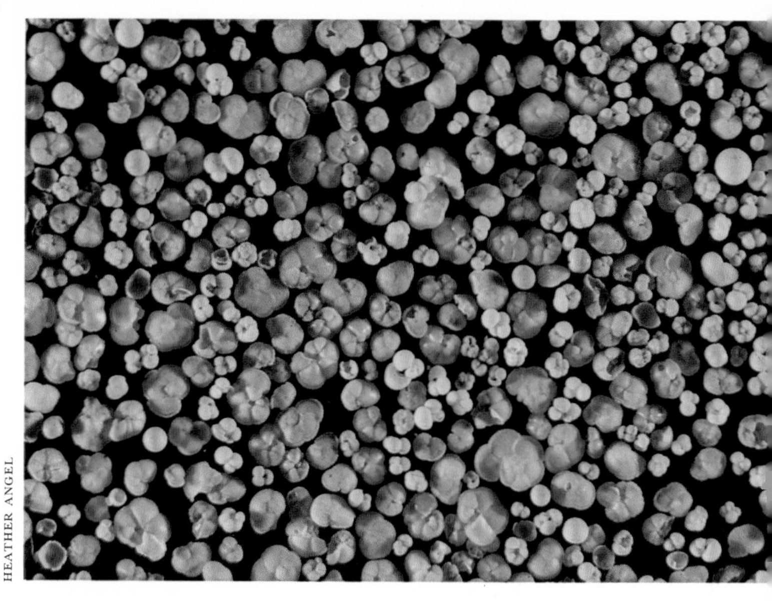

HEATHER ANGEL

Mud dredged up from 1,200 m in the West Atlantic, not far from the Cape Verde Islands. It is composed of the calcareous skeletons of seven species of Foraminifera. The largest of these single-celled planktonic animals is less than a millimetre across.

Right: Ecologically the Antarctic convergence (AC) is the most important feature in the current patterns of the South Atlantic. The cold surface water flowing northward from the ice edge is overridden by the southward flowing warm water. The cold water, Antarctic Intermediate water (AI), flows on north at depths of about 1,000 m to well beyond the equator. The South Atlantic bottom water (SABW) is formed at the ice edge by cold water sinking down and flowing slowly northward along the sea-bed. In the surface layers, a great anti-clockwise gyre dominates the tropical and subtropical latitudes, while in the region of the Antarctic convergence the winds of the Roaring Forties drive the massive westerly circumpolar currents of the West Wind Drift.

Newly formed ocean floor. Lava in the bottom of the median rift valley of a mid-oceanic ridge has solidified into the highly folded pahoehoe pattern and is completely free of any covering sediment. A compass hung below the camera helps to orientate the picture.

WOODS HOLE

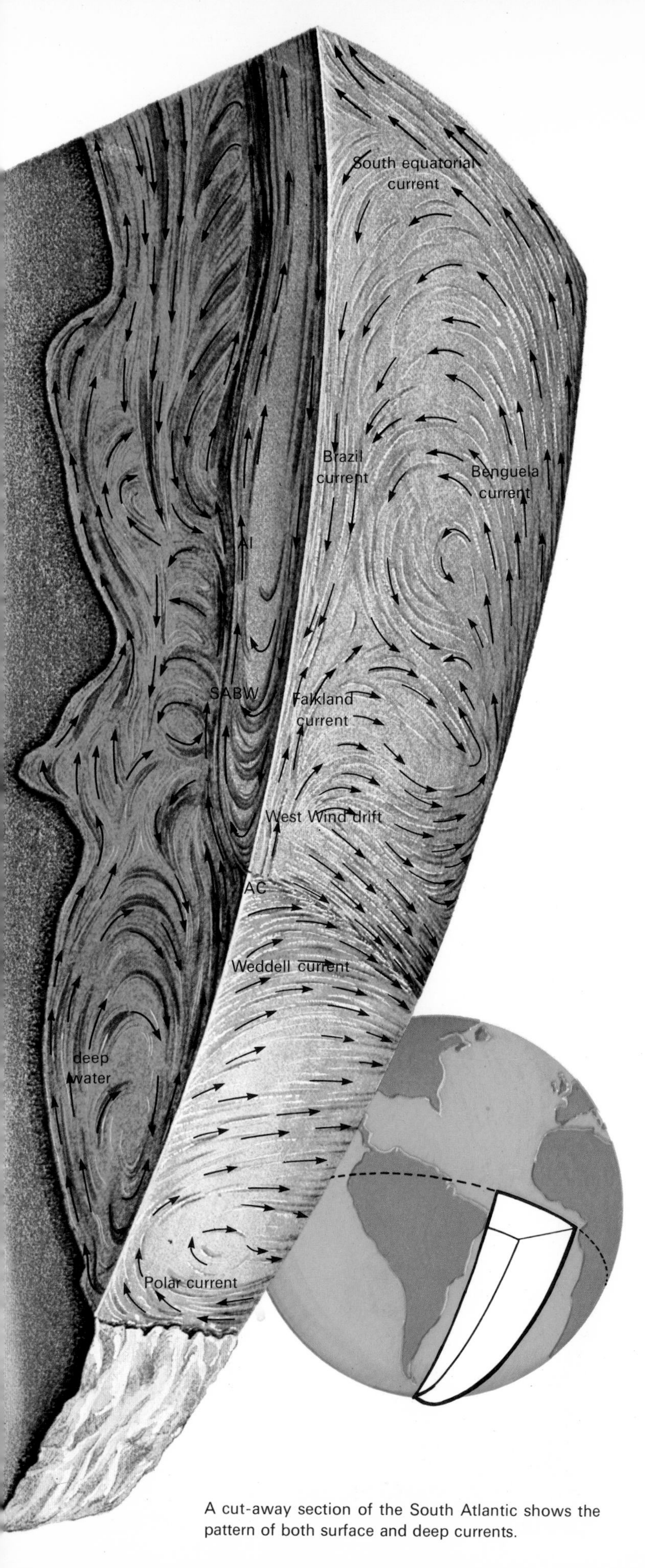


A cut-away section of the South Atlantic shows the pattern of both surface and deep currents.

biologically. Iron is an essential part of chlorophyll, a substance common to all green plants. Copper is a component of haemocyanin, which is the respiratory pigment of crustaceans and functions as an oxygen carrier in their blood, in the same way as does haemoglobin in human blood. Vanadium is concentrated hundreds of thousands of times by salps to be incorporated into their respiratory pigments. Manganese and cobalt are used in the organic compounds involved in the breakdown of sugars to give energy in both animal and plant respiration. The ability of certain animals to concentrate elements from their very dilute concentrations in sea-water, can make the disposal of toxic wastes into the sea potentially dangerous. A common little copepod, *Acartia*, that is abundant both in estuaries and in the open ocean, concentrates lead up to several hundred times, a potential hazard to predators.

Minute traces of organic compounds occur dissolved or suspended in sea-water. Their role is not fully understood but some act as poisons to inhibit the growth of competitor species, others encourage the growth of phytoplankton, the microscopic plants of the ocean. Many organic compounds leak away into the water when carnivores eat their prey (as much as a quarter of the body weight is lost from a prey copepod being eaten by a prawn) and can provide a source of food for bacteria or even quite large animals which 'fish' with mucus sheets onto which organic molecules may stick chemically.

Even more important biologically are the amounts of nutrient compounds dissolved in the water. Nitrates and phosphates are basic requirements for the growth and reproduction of the phytoplankton and silicates are also needed for some types of phytoplankton to grow.

## Water masses

Although the relative concentrations of the chemical ions vary little in sea-water, with the exception of the nutrient salts, the overall amount of salt dissolved in sea-water varies geographically and with depth. In most regions each kilogramme of sea-water contains 33–35 grammes of salts, but in the Red Sea levels of 40 grammes per kilogramme occur. (The notation °/₀₀ is usually used to express these values, e.g. sea-water in the Red Sea has a salinity of 40 °/₀₀.) Sea-water becomes heavier (i.e. denser) if it is either cooled or becomes more salty and, conversely, if it is warmed or diluted, it becomes lighter or less dense. At some latitudes more water evaporates from the sea surface during the year than falls on it as rain. The surface sea-water at these latitudes becomes saltier and heavier, and so tends to sink under water from neighbouring latitudes.

When ice forms, salts are left behind in solution, making the surrounding waters progressively more saline. During winter in the Antarctic this very cold and very saline water, rich in dissolved oxygen, sinks right down and forms the bottom water of all the major southern oceans. This source of oxygen to the deep ocean makes life possible at the greatest depths. When the ice melts in the Antarctic summer, it both cools and dilutes the surrounding sea-water, which flows slowly northwards until it encounters warmer but more saline surface water at the *Antarctic Convergence*, a region of great importance in the ecology of the Antarctic. Although it is more dilute, the colder water is denser and so sinks, forming a layer known as the *Antarctic Intermediate Water*. This layer flows on northwards at a depth of about 1,000 m and can be identified by its lower salt content well north of the Equator.

A great deal of information can be discovered from the temperature and salinity properties of the sea-water at a given position. The bodies of water with such characteristic properties are called *water masses* and the origins of most are now quite well known. For example, the Red Sea and the Mediterranean are inland seas with narrow sill entrances. Few rivers run into either and so much more water is lost by evaporation than is replaced by rainfall or run off. In the inner parts, the water therefore gets progressively saltier and so denser. Eventually this saltier but warm water sinks. There is a strong surface inflow through the Straits of Gibraltar which used to cause great problems to sailing ships trying to leave the Mediterranean. Underneath the inflow there is a smaller but significant outflow. This outflow of water, known as *Mediterranean Water*, is of relatively high salinity and can be traced over a wide area of the North-east Atlantic at depths around 1,000 m.

## Ocean currents

Currents would be produced just by the variations in density which result from geographical differences in the amount of rainfall and the freezing and melting of sea-ice. However, other powerful forces generate and modify the current patterns. There is a persistent pattern of winds, the trade winds, that blow across the oceans. The friction of these on the surface not only generates waves, but also pushes the surface water along as a current. The rotation of the Earth imparts a spin to the water movement so that immense *gyres* are formed, clockwise ones in the northern hemisphere and anti-clockwise in the southern hemisphere. Very fast currents flow northwards up the western boundaries in the northern oceans and up the eastern boundaries in the southern hemisphere.

In the North Atlantic, the Gulf Stream meanders and breaks up into a series of swirls and eddies off the eastern coast of North America, and continues as a much broader, slower flowing current named the *North Atlantic Drift*. The warmth of this current improves the climate of Western Europe. There is evidence that cycles of high and low atmospheric pressure cause fluctuations in the strength of the North Atlantic Drift, so giving rise to eleven year cyclic changes of mild followed by cooler winters.

A gentle breeze blowing over the sea's surface causes the formation of windrows parallel with the wind's direction. In the Indian Ocean these windrows are marked by aggregations of the blue-green alga *Trichodesmium*, an alga important in the ecology of tropical oceans.

HEATHER ANGEL

One part of the North Atlantic Drift flows on up into the Norwegian Sea, while another turns south, flowing down towards the African coast as the *Canaries Current*. Off the coast of Senegal, the Canaries Current swings westwards as the *North Atlantic Equatorial Drift* back across the Atlantic.

Similar gyres occur in each half of the major oceans, although greatly modified in the Indian Ocean by the seasonal changes of the monsoon winds. Each gyre has its own typical assemblage of drifting animals and plants, and they play an important role in the migration of many of the larger fish.

Apart from the fast flowing boundary currents, most currents are a complex of swirls and eddies of all sizes, which average out as a general flow in some direction. The complexity of these eddy patterns can be appreciated by putting one or two drops of ink in a bath full of cold water and seeing the effects of running in a little warm water.

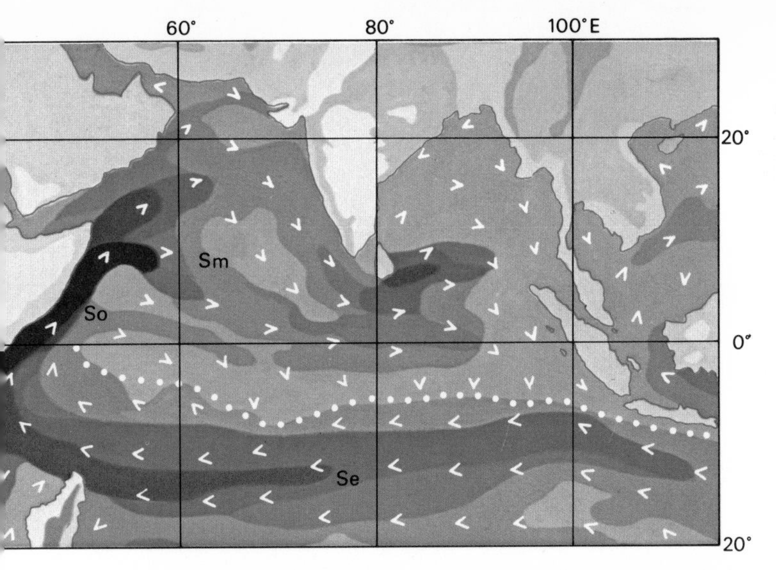


Patterns of surface currents in the Indian Ocean reverse during the northern hemisphere summer as a result of the change in the direction of the persistent monsoon winds, which occur every six months. The along-shore winds cause upwelling along the Somali and South Arabian coasts during this regime of wind and current. During the other monsoon, upwelling occurs along the western coast of the Indian subcontinent.

**So Somali current. Sm South West Monsoon current. Se South Equatorial current.**

Patterns of the surface currents in the world's oceans during the northern hemisphere winter, showing the average speed and direction of the various currents.

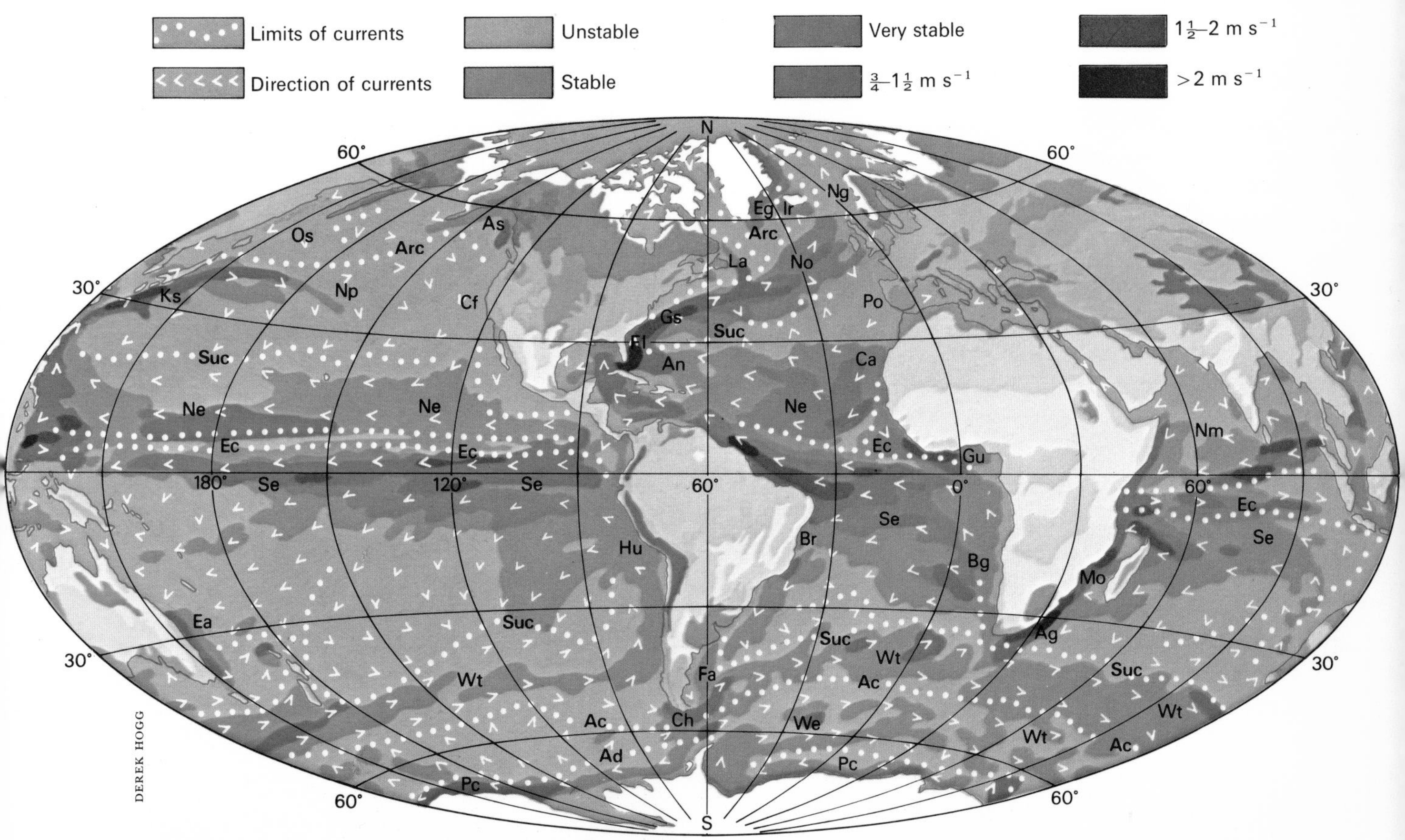


**Ag Agulhas. An Antillean. As Alaskan. Bg Benguela. Br Brazil. Ca Canary. Cf California. Ch Cape Horn. Ea East Australian. Ec Equatorial Counter Current. Eg East Greenland. Fa Falkland. Fl Florida. Gs Gulf Stream. Gu Guinea. Hu Humboldt or Peru. Ir Irminger. Ks Kuroshio. La Labrador. Mo Mozambique. Ne North Equatorial. Ng Norwegian. Nm North East Monsoon. No North East Atlantic. Np North Pacific. Os Oyashio. Pl Polar. Po Portuguese. Se South Equatorial. We Weddell. Wt West Wind Drift.**
**Arc Arctic Convergence. Ac Antarctic Convergence. Ad Antarctic Divergence. Suc Subtropical Convergence.**

## Wave formation

Dropping ink into a bath will also illustrate some of the simpler phenomena of waves. From where the drops hit the surface, series of little waves or ripples radiate outwards. As they spread out, the ripples not only get smaller but are also further apart. When the ripples hit the side of the bath they are reflected back, and two trains of ripples can be seen travelling across each other. The drop falling on the water surface imparts energy to the surface, which is dissipated in the form of waves. In the ocean, the friction of a wind blowing over the surface of the sea imparts energy to the surface, producing waves. The stronger and more persistent the wind, the larger are the waves produced. The waves radiate out from a storm, gradually becoming lower and with the crests progressively further apart. Sailors call these swell, as they often cause much more uncomfortable rolling of a ship than the higher, but shorter, locally produced waves. The most uncomfortable conditions are when two or more trains of waves travel across each other, producing a 'confused' sea. Wind and waves play an ecologically important role in churning up the surface layers of the sea, but at depths below 100 m only in the most violent storms will the ocean's inhabitants experience any movement from surface waves.

SEAPHOT

A wave rolling inshore feels the drag of the shallows and curls over into a magnificent breaker. The pounding of waves on a sandy beach can make it uninhabitable for animals. On exposed rocky shores only a few specialized species can cling on.

## Thermoclines

In the tropics, the heat of the sun warms the surface layers and greatly reduces their density. This warm, lighter water floats on top of the deeper, colder, denser water. The boundary between them, where the temperature drops sharply with depth, is known as a thermocline. The sharp increase in the density of the water below the thermocline prevents any vertical mixing of the water across it. In fact the thermocline can behave much like the sea surface and is thrown into a series of waves by any disturbance. The ecological significance of the thermocline in the tropics is that the tiny plants living in the sunlit waters above it, quickly use up all the mineral nutrient salts while the nitrates and phosphates they also need tend to drop down below the thermocline in the faecal pellets of the animal grazers. Thus the growth of the phytoplankton is limited and the productivity of tropical waters is much lower than that of higher latitude areas.

In temperate latitudes thermoclines are seasonal. One is formed in the spring as the weather begins to

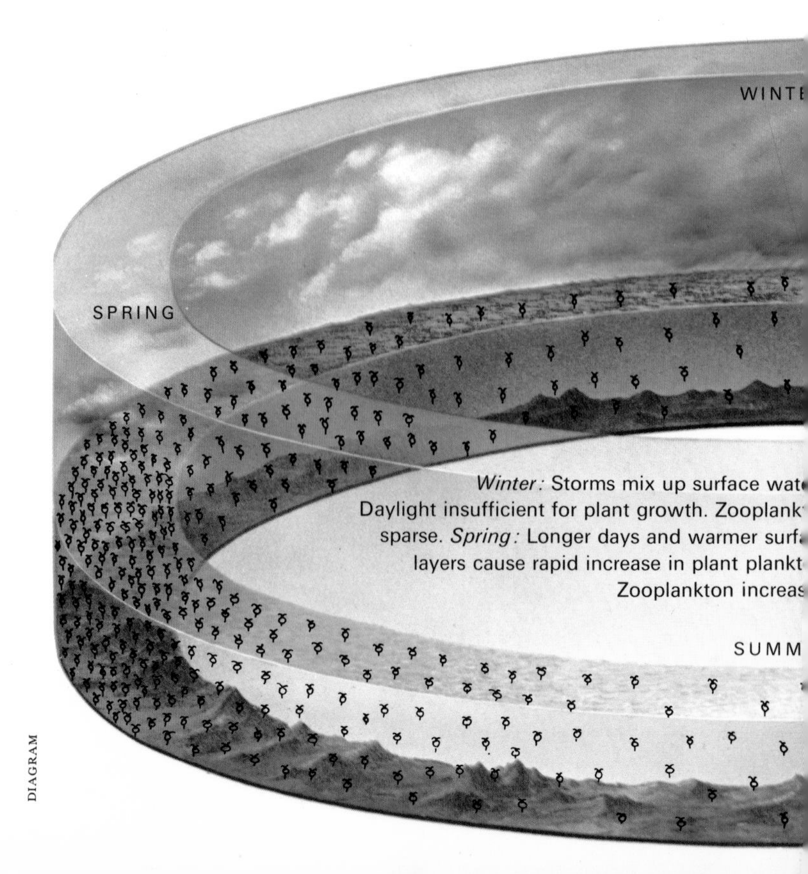


DIAGRAM

*Winter:* Storms mix up surface wat
Daylight insufficient for plant growth. Zooplank
sparse. *Spring:* Longer days and warmer surf
layers cause rapid increase in plant plankt
Zooplankton increas

get warmer and it is overturned by the autumnal storms. During the winter the sea is the same temperature to a much greater depth and there is considerable vertical mixing, allowing the mineral nutrients to be carried back up into the surface layers. Phytoplankton growth is inhibited during the winter because the daylight hours are short and there is little bright sunlight. The springtime increase in day length stimulates a great blooming of the phytoplankton, usually coinciding with the beginning of stratification—the formation of a new thermocline. The zooplankton, the tiny animals which graze on the phytoplankton, respond to the bloom of the plants by growing rapidly and reproducing and within six to eight weeks the phytoplankton bloom is in decline because so many animals are grazing on it. Meanwhile the formation of the thermocline means that the levels of the mineral nutrients in the surface water are being reduced. Without nutrients the phytoplankton cannot grow and reproduce fast enough to outstrip the feeding rate of the animals, and it gradually declines until the autumnal turnover. Then there is a short bloom which peters out as the day length shortens once again and bright sunlight is rare.

## Upwelling

Despite the permanence of the thermocline in the tropics in certain areas, cold deep water from below may be churned up to the surface either by the pattern of the ocean currents, or by a combination of the effects of offshore winds and the rotational (Coriolis) forces generated by the Earth's spin. This deep water, from 100–200 m down, carries with it an abundance of nutrients and the phytoplankton grows so fast that the sea becomes discoloured. These 'blooms' are followed by a great increase in the zooplankton either because of a higher breeding rate or because more animals move into the area to feed on the immense resources of food available. The zooplankton in turn becomes the food of vast shoals of fish. These 'upwelling' areas are the centres of the world's great fisheries.

Off the coast of Peru, the cold Humboldt Current, flowing northwards up the South American coast, swings westward to become part of the *Pacific South Equatorial Current*. Upwellings occur at many places along the coast throughout most of the year.

Off the South Arabian coast, the upwelling is seasonal. During the south-west monsoon, it occurs at several places, often in the regions of capes or some of the offshore islands such as the Kuria

The seasonal cycle of zooplankton abundance at temperate latitudes. Two plant plankton blooms, the largest in spring and a smaller one in autumn bring corresponding increases in the zooplankton which feed on the tiny plants.

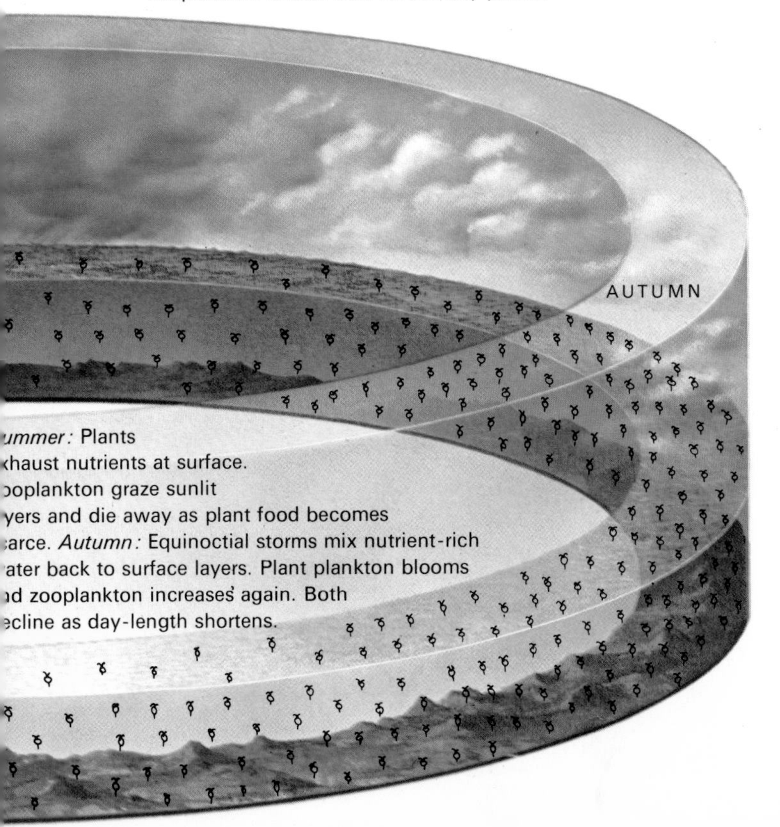

Murias. When the upwelling first occurs, warm-water fish are sometimes caught in the newly upwelled water, which may be as much as 12°C colder than the water it displaces; vast numbers are killed by the cold shock. As the upwelling continues, great blooms of phytoplankton occur, some of which may produce powerful toxins that not only cause large fish mortalities, but are also lethal to man if he eats contaminated shellfish. Some of these toxins, produced by types of phytoplankton called dinoflagellates, are amongst the most powerful known and act by interfering with the junction between nerves and muscles.

The very high productivity of the area off the South Arabian coast results in eutrophication; this is similar to what happens when a lake or estuary is heavily polluted with sewage. The large input of organic material into the water speeds up the growth of decomposer bacteria, which feed by breaking down the organic matter aerobically—i.e. by using the oxygen dissolved in the water to oxidize these compounds. This greatly increased demand for oxygen rapidly uses up the supply in the deeper water below 200 m. If this deoxygenated water is then upwelled, it suffocates the fish living in the surface layers almost immediately.

## Sulphide production

If all the oxygen is used up by this bacterial breakdown of the organic compounds, the growth of sulphur bacteria is stimulated. Sulphur bacteria break down the organic compounds anaerobically, not using free oxygen but oxygen produced by a chemical reaction in their cells. The hydrogen sulphide produced by this reaction not only has a revolting smell of bad eggs but is nearly as poisonous as hydrogen cyanide. Sulphide production is a natural phenomenon in rich muds of mangrove swamps and estuaries and is an important environmental factor in controlling the distribution of many of the worms and molluscs that inhabit these muds.

It is also a natural phenomenon in the Black Sea—called 'black' because iron lowered below 100 m is found to be covered with a black sulphide layer when it is raised again. The Black Sea is an inland sea with only a shallow connecting channel via the Bosporus into the Mediterranean. The surface layers are well oxygenated, but they are brackish, diluted by the freshwater runoff from the rivers. The density difference between the dilute surface layer and the deeper saltier water is too great to allow them to mix. Oxygen can get into the deep water by two methods only: firstly by vertical diffusion, which is far too slow to prevent the formation of the sulphide, and, secondly, by the bottom current up the Bosporus. This water coming from the Sea of Marmara is partially depleted of oxygen, and only flows fast enough to renew the bottom salty water of the Black Sea once in 2,500 years.

The Baltic is analogous to the Black Sea. It has a very narrow entrance from the North Sea, through the Skagerrak and Kattegat between Denmark and Sweden. The surface waters are brackish, diluted

LUBBOCK/PHOTO AQUATICS

In a shoal of sardines, each fish is orientated with reference to its neighbours. At night, when there are no visual cues, the regular structure becomes more chaotic and tends to disintegrate.

In the rich upwelled waters off the coast of Peru the shoals of anchovetta provide food for vast numbers of guanays. The droppings of these cormorants, deposited on off-shore islands, used to be the world's main source of nitrates.

DOSSENBACH/NSP

by the freshwater runoff from the rivers. In the Gulf of Finland the water is so dilute that freshwater animals live alongside those marine species that are well known as being tolerant of brackish water. The deep water of the Baltic has the same salinity as that of the North Sea. It is renewed about every twenty-five years, not by a continuous current, but by intermittent inflows of salt-water spilling over the sill into the deep inner basin. Scientists began to measure the oxygen content of the deep water at the beginning of the century in association with fishery investigations. The surface water was found to contain over 5 ml of oxygen dissolved in every litre, but the deeper water contained only 2 ml.

During the next fifty years, the amount of oxygen slowly decreased to about 1 ml per litre, a phenomenon probably associated with the growth in human populations near the Baltic. Then, quite suddenly, the oxygen content of the bottom water began to decrease more rapidly. A great social revolution had taken place: detergents had been discovered. Detergents contain large quantities of phosphates and these, emptying into the sea with waste water, stimulated the growth of micro-organisms. These, in turn, used up the oxygen more rapidly, with a disastrous effect on the fish.

NEVILLE COLEMAN

In the tropics, upwelling often induces the formation of dense blooms of plant plankton which discolour the sea and are sometimes highly toxic. This bloom of the blue-green alga *Trichodesmium* is harmless.

## Sunlight

Another important ecological factor that changes with depth is the amount and the colour of light that penetrates to any given level. A significant proportion of sunlight is reflected back from the sea's surface and so never penetrates it at all. This proportion increases as the sun becomes lower in the sky and also if the sea surface is roughened by bad weather. The water itself absorbs light, absorbing the longer wave lengths, i.e. the red end of the spectrum, more rapidly than the shorter wave lengths. Any underwater photographer will have found that, even in very shallow water, colour pictures have a decidedly green cast. This is because the very rapid absorption of the red light upsets the colour balance. Even in the clearest water, all the red light has been absorbed by the time a depth of 30 m is reached. Since red-coloured objects look red because they reflect red light and absorb the

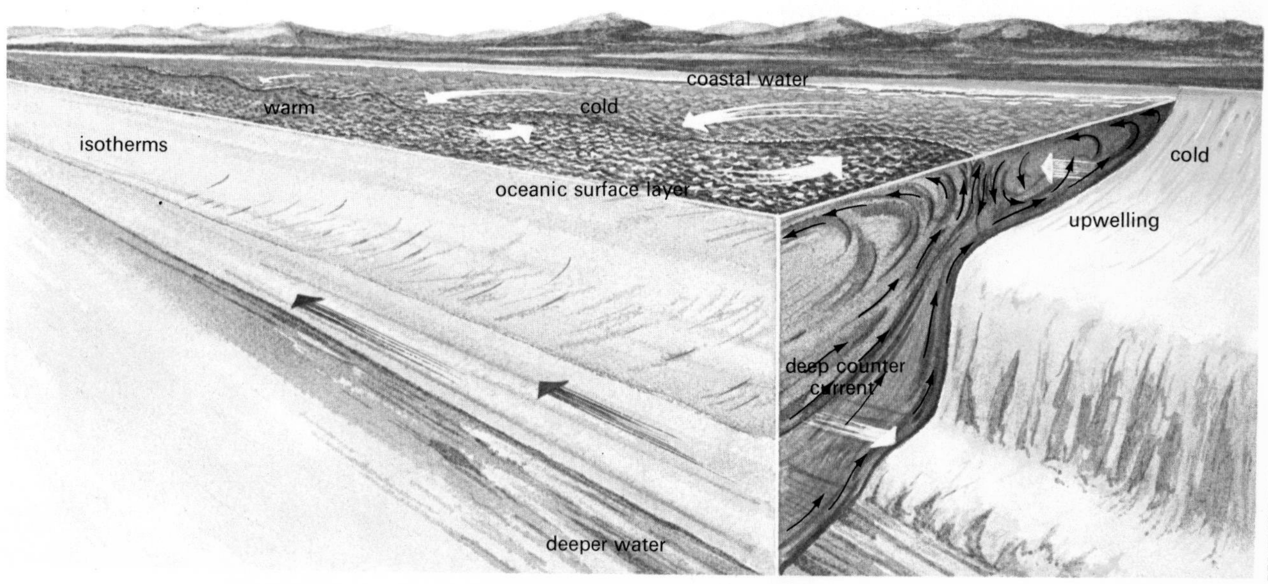


A complex pattern of currents occurs in the upwelling region off the Namibian coastline. Winds blow north-westerly up the coast, pushing the surface waters along. The earth's rotation causes the water to swing to the west and move off-shore. It is replaced by cold water welling up from depths of 100–150 m carrying with it mineral nutrients, which induce blooms of phytoplankton.

BRIAN PRICE-THOMAS

other wave lengths, below 30 m, where there is no longer any red light to reflect, they will look black.

Sunlight is the initial source of energy for all life on Earth. It is used by green plants to synthesize energy-rich compounds, such as glucose, from the simple molecules of carbon dioxide and water by the process of photosynthesis. The green pigment chlorophyll plays an important role in this manufacturing process, and is present in all photosynthetic organisms; even those such as the brown fucoid seaweeds, which cover rocky shores in temperate latitudes and in which the colour of the chlorophyll is masked by other pigments. The plants also build up, from the simple building blocks of amino-acids and carbohydrates, more complex structural molecules such as proteins and cellulose. For plants to grow, the amount of the sun's energy they store during daylight hours must exceed the amount they expend throughout the day and night in respiration. Animals cannot synthesize energy-rich substances from simple molecules as plants do and so they all ultimately depend on plant material as their source of energy, even if the material is second, third or fourth hand.

Since light is progressively more absorbed with increasing depth, a depth is reached where the amount of energy from sunlight which the plants can utilize exactly balances their respiratory requirements. This depth is called the *photosynthetic compensation depth*. Above the compensation depth is the *euphotic zone* where all plant growth occurs; all organisms that live permanently below it have to rely on the rain of organic material sinking from the euphotic zone above. In very clear oceanic water the photosynthetic compensation depth is at about 100 m. This is roughly the depth at which the intensity of light has been reduced to about 1 % of the intensity at the surface. However, the photosynthetic compensation depth is much shallower if the water is very turbid, either because of suspended detritus or because of an abundance of phyto- and zooplankton. It will also be shallower when more light is reflected, when the water surface is very rough, or if the sun is low in the sky.

When sunlight passes through the sea surface, the refraction of the light at the surface causes a circle of brightness which is called *Snell's circle*. If the sun is low in the sky the circle is distorted into an ellipse. Thus it is possible for an organism below the sea surface to perceive the elevation of the sun above the horizon. However, with increasing depth, the scattering effect of the water renders Snell's circle progressively more indistinct. At about 250 m

GRUHL/PHOTO AQUATICS

Red light is rapidly absorbed by sea-water, so by naturally occurring light both the fish and the Gorgonian corals seen here will appear black. It is only when a photographer uses a flash-light that the superb colouring of these animals is revealed.

Snell's circle, the sharp-edged arc of light penetrating through the water, is caused by the refraction of light at the surface. The dourado or dolphin fish is silhouetted against the light, making it vulnerable to predators below.

GRUHL/PHOTO AQUATICS

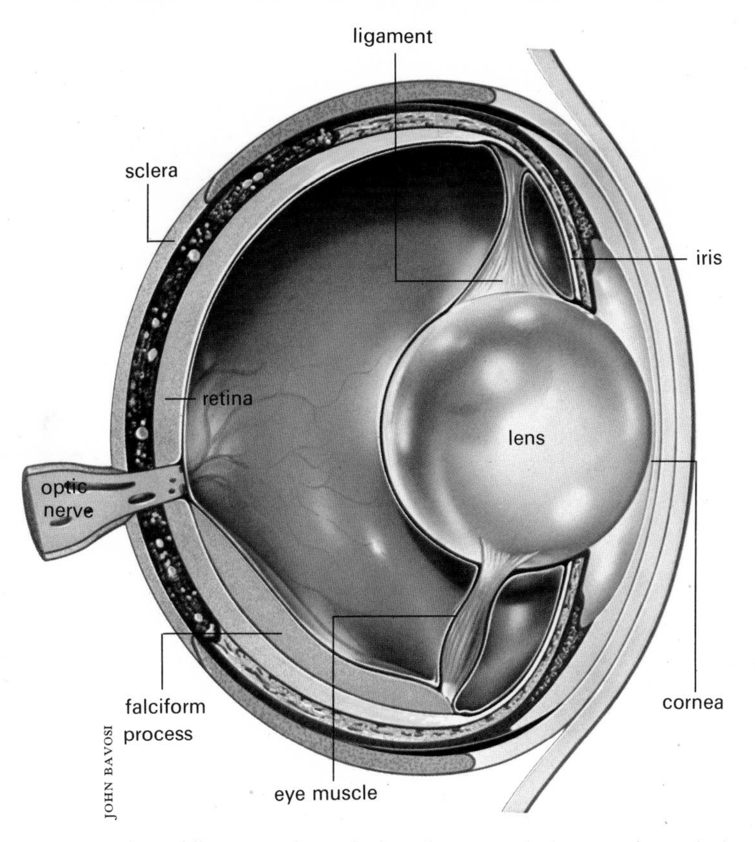


A semidiagrammatic vertical section through the eye of a typical bony fish showing its major features. The circular lens gives the fish almost completely all-round vision—a considerable advantage in an environment in which danger may approach from any direction, vertically as well as horizontally.

Many deep-sea fish have elaborate eyes, adapted to look for their prey silhouetted against the relative brightness of the down-welling light. *Benthabella* has tubular, upward-looking eyes and a special lens pad which allows it to see sideways.

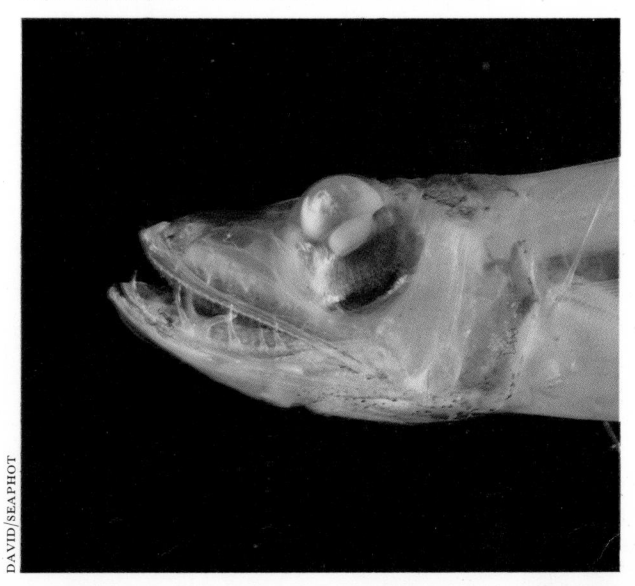


depth, all evidence of the sun's elevation is lost and Snell's circle can no longer be seen. The distribution of light intensity is then quite symmetrical, with the brightest light coming from directly overhead. The dimmest light is scattered back from the depths below. As we shall see, this boundary of 250 m has great ecological significance.

Light continues to be absorbed with increasing depth. In the clearest water, the deepest depth at which the human eye can perceive daylight is 800 m but deep sea fish have eyes specially adapted for seeing in very low intensities. Elaborate lens systems collect the maximum amount of available light. While human retinas contain two types of light sensitive cells, rods, and cones (the latter are used to observe colour), a fish's retinas contain only rods. The fish's retinas also contain a much greater density of the visual pigment rhodopsin than the human eye. Thus fish can see daylight at depths of 1,000 m or even 1,300 m in exceptionally clear oceanic water. At greater depths, there is insufficient daylight left for even these specially adapted organisms to perceive.

The changing colour balance of the penetrating daylight has important repercussions when the function of animals' colours is considered. Only in shallow clear seas, such as over coral reefs, does the colour balance of the penetrating daylight allow elaborate colour patterns to be of significance.

## The influence of tides

Tides are ecologically extremely important in the fringing seas. Tides are generated by the gravitational forces of the moon and sun pulling out the waters of the ocean. The moon's axis of rotation passes approximately over the equator. Water on the side of the Earth nearest the moon is pulled out by its gravity into a high tide. On the opposite side away from the moon water also piles up because there the moon's gravitational pull is reduced, so there are two high tides produced by the moon. Since the moon takes 24 hours 50 minutes to encircle the Earth there is a semi-diurnal lunar tide every 12 hours 25 minutes. It is interesting to note that tides also occur in the Earth itself with amplitudes of about 1–1½ metres. The Earth's axis is inclined towards the sun so that the sun pulls out one tidal bulge in the northern hemisphere and another in the southern. Sun induced tides are therefore diurnal. In some areas the configuration of the tidal basins is such that the resonance of the lunar semi-diurnal tides is damped and the main tidal rhythm is then the sun-induced diurnal tides. A further effect of the sun is to modify the ranges of the semi-diurnal lunar tides into springs and neaps.

In the deep ocean, tides conform quite closely to a simple basic pattern. However, a high tide behaves very much like a wave: it becomes modified as it moves into shallow water and its oscillations may be dampened or magnified by the shape of a tidal basin, or by a tidal wave coming from another direction. For example, a high tide moving round an asymmetrical island will take longer to travel round one side than the other. This may cause strange tidal phenomena, such as double high tides, or long periods of a tidal stand when there is neither a rise nor a fall. Estuaries may have a funnelling effect which magnifies the tidal range, as in Chesapeake Bay on the east coast of the United States. In some estuaries, the tide builds up into a real wave or bore which rushes up the tidal part of the river. Similarly, violent storms coinciding with high spring tides, may push up extra high tides.

In contrast, some regions experience very small tides. The Mediterranean and the Baltic have very small tidal ranges because they are too small to have locally generated tides of any significance and the entrance straits are too narrow to allow much tidal effect from the outside ocean. Other places are positioned at the nodes of tidal oscillations. If water is set rocking in a dish, there will be no change in depth along the axis of the movement of the water. A similar effect occurs with the ocean oscillating in a tidal basin.

On shores with wide-ranging tides, a large, special intertidal fauna and flora develops. These are organisms that can withstand a progressively increasing amount of exposure as they live higher and higher up the shore towards the high tide line. They not only have to be able to withstand the drying effects of the wind and the heat of the sun, but also flushing with freshwater by rainstorms during low tide or even exposure to extreme cold. In Antarctica there is no intertidal life because of the scouring effects of the pack ice and icebergs.

Even below the low tide line, tidal currents have important modifying influences on the animals that live there. Rocks are scoured free of mud; detritus is swept along by the currents, carrying food to many of the inhabitants; sand and gravel are swept into huge waves which, like sand dunes, slowly move across the sea-bed. The currents also help to disperse the planktonic larvae of both shore and sublittoral animals. Tides are a dominant influence in the ecology of the majority of organisms living on the continental shelf. In deep oceanic water, tides still occur but pass unnoticed by casual observers.

HAWKES/NHPA

Along the length of Chesil Beach, England, the lines parallel to the sea mark successive high tide marks, the highest left by the last spring tide. As the tides progress towards neaps, each high tide is lower than its predecessor.

## The ocean environment

The oceanic environment lacks the complex organization into numerous microhabitats that occurs in most terrestrial environments. However, the ocean is not a uniform environmental bath; its complexity is derived from a network of environ-

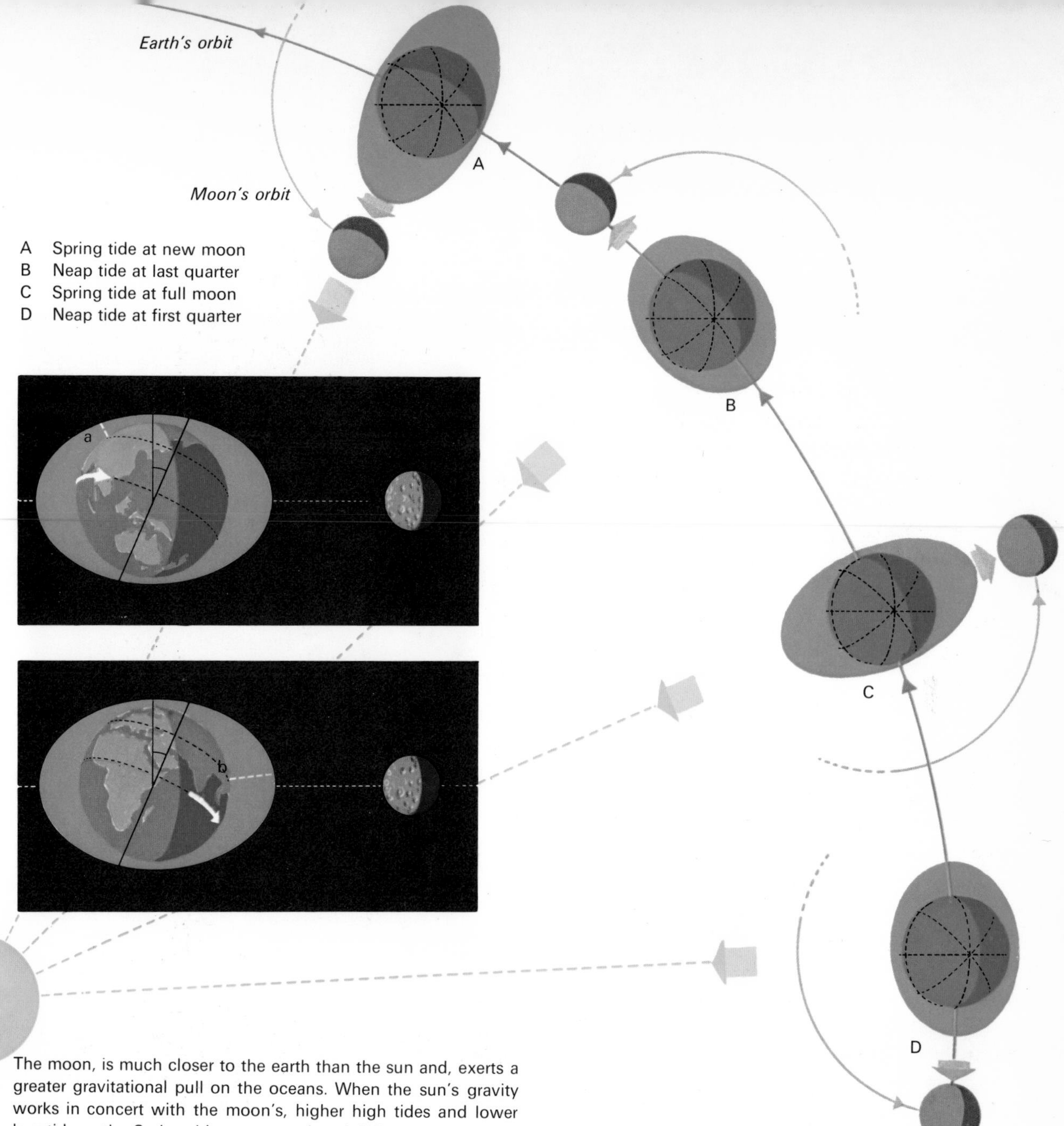


The moon, is much closer to the earth than the sun and, exerts a greater gravitational pull on the oceans. When the sun's gravity works in concert with the moon's, higher high tides and lower low tides—the Spring tides—are produced. When the gravity of the sun is working antagonistically, Neap tides occur, with lower high tides and higher low tides.
The moon passes almost directly over the equator and produces approximately two high tides and two low tides each day, i.e. a semidiurnal tidal cycle. The earth's axis is inclined at an angle of 23·5° to the ecliptic (the plane of the earth's orbit round the sun). At a given latitude, at position (a), the sun produces a low tide; twelve hours later at position (b), the sun produces a high tide. So because of the earth's tilt, sun-induced tides are diurnal.

mental gradients. Mid-water communities grade almost imperceptibly into each other. On the bottom the variations in the structure of the sea-bed produce more complex communities. This complexity increases in shallower water and reaches its greatest degree in the tropical coral reef communities, where the coral itself has a heterogeneous three-dimensional structure that is similar in complexity to a rich forest environment.

Richer and more varied communities occur in environments that are relatively constant. When the environment is subject to change, animals living there have to expend greater amounts of energy in order to survive, and this appears to limit the number of species that can occupy a habitat. Thus between the tides there are fewer species than in the more constant environments below the low tide line. At high latitudes, where there is marked seasonal variation in the light and the weather conditions, fewer species occur than in comparable habitats in tropic latitudes where these seasonal variations are much smaller. Geological history is also important in determining the richness of communities; the Indo-Pacific coral reefs are older geologically and far richer in species than the younger Caribbean coral reef faunas.

# The Ocean's Fishes

Fishes are the vertebrates that dominate the Earth's seas, occurring on shores and down into the abyssal depths, in the coldest polar waters and the warmest parts of the Red Sea and Persian Gulf. They first occur as fossils in the Silurian era in freshwater deposits laid down 400 million years ago. This early evolution in freshwater has resulted in all fishes having blood containing salt at about half the concentration of that in sea-water. Not only do all fishes have blood of this concentration but so do all their descendants, including man.

## Jawless fishes

Most of the earliest fish which occurred in the Silurian and Devonian eras belonged to groups such as the cephalaspids, which had heavy armoured heads, and the anaspids, which were small fishes covered with bony scales. All these early fishes had one common feature; they were jawless. Although most of them have long been extinct, one group of fishes still survives—the cyclostomes. These include the lampreys, some of which undergo spawning migrations analogous to those of the salmon. They migrate up rivers to spawn and there spend their early larval life. The larvae descend the rivers to the sea where they spend a few years maturing before returning to breed. The other cyclostomes are hagfish (slime-eels), which are abundant scavengers on the bed of the deep ocean. Camera systems used by scientists to film very deep living bottom fishes attracted to baits often picture writhing masses of the slime-eels. Both lampreys and hagfish have sucker mouths which rasp away at dead or live flesh. They have a special arrangement of muscular gill pouches by means of which they can pump water through the gills and still maintain a grasp on their prey. Even large whales bear circular scars showing where an attacking cyclostome has twisted out a core of flesh.

## Elasmobranchs

Late in the Devonian era the first of the elasmobranchs occurred. Like the jawless fish, they have cartilaginous skeletons that are not calcified into bone. They are a diverse group and include animals ranging from the sluggish plankton-feeding whale-shark to the most feared man-eater, the great white shark, *Carcharodon*. There are about 250 species of sharks in all. The squaloids are sharks with a spine on the front of each dorsal fin; a typical member

HEATHER ANGEL

Lampreys are the present-day survivors of the once abundant group of jawless fish. The mouth of *Petromyzon marinus* is modified into an efficient sucker, armed with rasping teeth. There are seven pairs of gill slits down the side of the head.

NEVILLE COLEMAN

The ornate wobbegong shark (*Orectolobus ornatus*) at 10 m in Byron Bay, New South Wales. By day these sharks lie up on the bottom, the fringes on the head camouflaging them by obscuring their outline. At night they actively hunt their prey.

Pages 32–33
The saupe (*Sarpa* sp.), a shallow-water species of sub-tropical seas off the east Atlantic and south Indian Ocean, feeds by scraping microscopic algae off rocks and large seaweeds. They habitually swim in tightly packed shoals.

Right: A white-tipped reef shark (*Triaenodon apicaulis*) glides over a Queensland reef in only 2 m of water. The under-slung mouth, the broad-based rigid fins, and the long projecting upper lobe of the tail fin, are typical shark features.

is the spiny dogfish, *Squalus*. More unusual squaloids are the saw-sharks, *Pristiophorus,* the monk-fish, *Squatina,* and the thresher-shark, *Alopias*. Threshers are reputed to hunt in packs using their long whip-like tail fins to herd together shoals of fish on which they feed.

There is a group of six- and seven-gilled sharks whose fossil record extends back 150 million years into the Jurassic era. The cow-shark, *Hexanchus,* is the most abundant representative of these. It commonly reaches lengths of over 5 m, when it weighs about 1,300 kg. The size record was a monster of over 9 m caught off south-west England. They are amongst the most abundant sharks at depths of about 1,000 m, particularly in the Mediterranean.

Skates and rays are elasmobranchs which are specialized for life on the sea-bed. Instead of swimming by beating the tail as other fish do, they propel themselves with rippling movements which pass backwards along their greatly enlarged pectoral fins. Most feed on invertebrates grubbed up from the bottom, but an obvious exception is the manta ray which has become a plankton-feeder. Rays show interesting adaptations of their body structures, ranging from the powerful stings of the sting-ray, *Trygon,* to the modification of muscles to form an electric organ in *Torpedo*, the electric ray. There are about 350 species of skates and rays.

One aberrant group of elasmobranchs, which consists of only about twenty-five species are the rabbit-fish (chimaeroids). They have large-toothed mouths with a tiny aperture surrounded by broad lips. They were once widespread and abundant, but now occur mostly in fairly deep water.

## The ancestors of bony fish

Also late in the Devonian era the first ancestors of the bony fishes, the paleoniscoids, appeared. Some of these ancestral types survive today in freshwater species of paddle-fish and also the sturgeon, which migrates from the sea into rivers to spawn. Other ancestral groups include the Dipnoi, which have modern descendants in the lung-fish, and the Crossopterygii, which gave rise to the fascinating marine living fossil, the coelacanth. Described originally from a single specimen, caught in 1938 by a trawler off East London in South Africa, the coelacanth was immediately recognizable by its resemblance to fossils from the Carboniferous era. Since then, many more specimens have been caught

NEVILLE COLEMAN

Sharks of various types occur throughout the world's oceans, but in greatest diversity in tropical seas. Although the largest of all, the Whale shark is a plankton feeder, most sharks are fearless predators. The largest carnivorous shark is the Great White. A specimen of this shark over 11 m long was trapped in a herring weir off New Brunswick, Canada and is the largest ever accurately measured. Teeth from Miocene deposits about ten million years old suggest that this shark's ancestors may have grown to over 24 m in length.

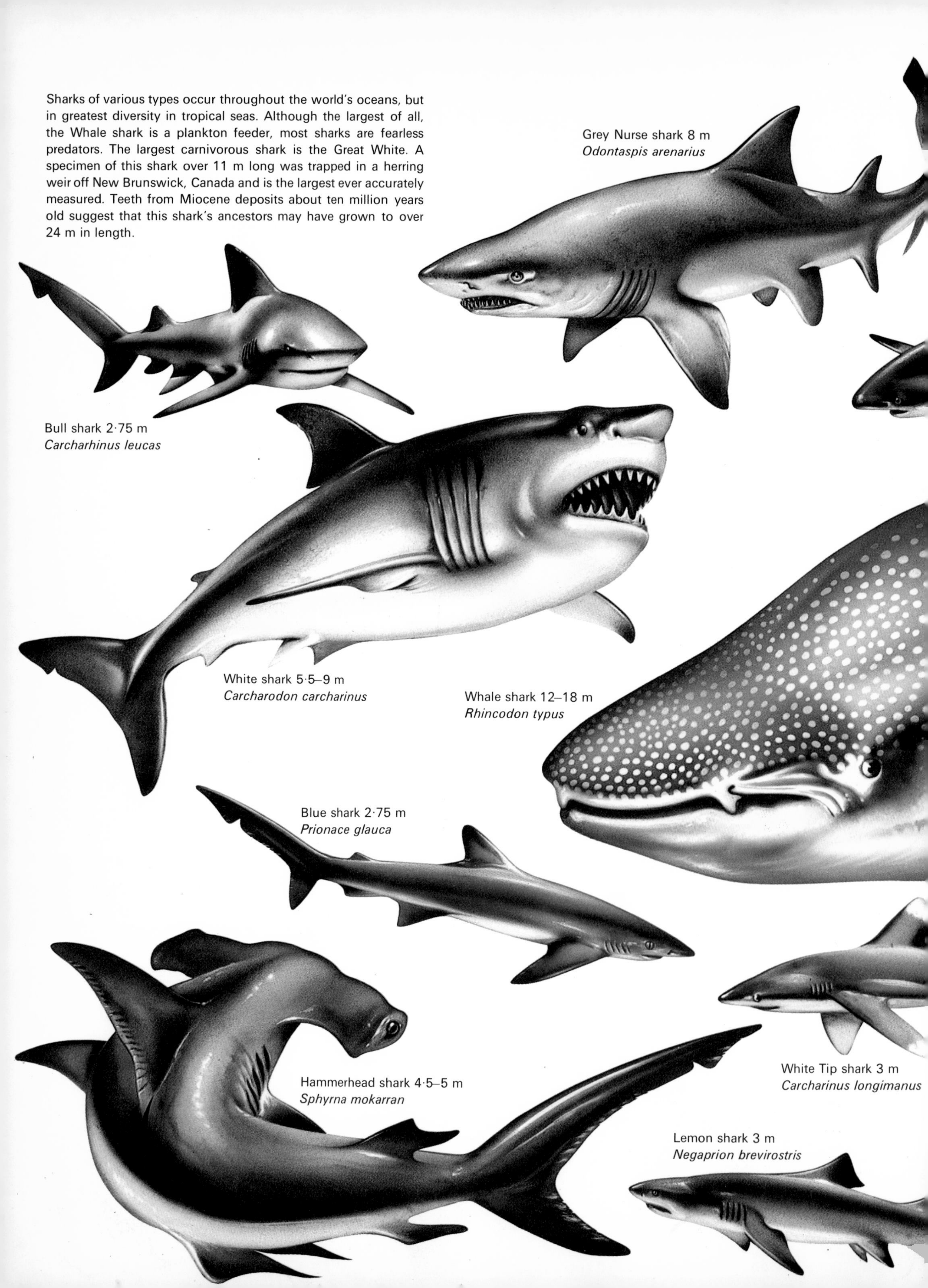

Grey Nurse shark 8 m
*Odontaspis arenarius*

Bull shark 2·75 m
*Carcharhinus leucas*

White shark 5·5–9 m
*Carcharodon carcharinus*

Whale shark 12–18 m
*Rhincodon typus*

Blue shark 2·75 m
*Prionace glauca*

Hammerhead shark 4·5–5 m
*Sphyrna mokarran*

White Tip shark 3 m
*Carcharinus longimanus*

Lemon shark 3 m
*Negaprion brevirostris*

Nurse shark 3–3·65 m
Ginglymostoma cirratum
The sharks are shown approximately to scale with one another and with the human figure.
Thresher shark 4·5–5·5 m
Alopias vulpinus
Sand Tiger shark 3–4·5 m
Odontaspis taurus
Mako shark 3·5–4·25 m
Isurus paucus
Leopard shark 2 m
Holohalaelurius regani
Tiger shark 4·5–6 m
Galeocerdo cuvieri

by native fishermen, using droplines, off the steep rocky volcanic Comoro Islands. The one known species, *Latimeria*, is of interest to scientists not only because it is a living fossil, but also because it is thought to be close to the ancestral stock that evolved into land vertebrates. Recently a live specimen was observed for the first time by biologists. It used its dorsal and anal fins to scull through the water, moving its paired pectoral and pelvic fins rather like limbs to manoeuvre it over the bottom. A museum specimen dissected in the U.S.A. has revealed that its large orange-sized eggs are internally fertilized and the young are born live. This, as we shall see below, is unusual in bony fishes, though not in sharks.

From the paleoniscoids arose the holosteans which were the dominant fishes during the Jurassic and early Cretaceous eras, 190–100 million years ago. The holosteans had thick heavy tooth-like scales, and survive today as the garpike and bowfin of the American Great Lakes. This group was gradually pushed out by the teleosts or true bony fishes which evolved from them.

## Teleost fish

With the exception of sharks and rays, virtually every fish seen in aquaria, on fishmongers' slabs, in the sea and in freshwater, is likely to be a teleost. There are at least 20,000 species, which include an amazing variety of forms adapted to different types of life-styles. The basic teleost body is a spindle shape, streamlined for smooth swimming through water. There are various fins: one or two dorsal fins down the back, a tail (caudal) fin, and on the underside a pair of pectoral fins, usually just behind the gill covers, and, further back, a pair of pelvic fins and an unpaired anal fin. The fins are basically used for swimming and trimming the fish's position in the water, but they may be modified for a great variety of other functions. In the gaudily coloured lion-fish, *Pterois*, the striped fins advertise not only the fish's dangerous nature but also that the spines themselves are poisonous and other animals should keep away from them.

The mudskippers, *Periophthalmus*, which frequent the edges of mangrove swamps, use their dorsal fins in sexual display. In remoras, which hitch rides on sharks or other large fish, the dorsal fin has been modified into an effective sucker organ. In flying fish, the pectoral and pelvic fins are expanded into great wing-like structures which can be clicked out into position for their gliding flights.

KOPP/PHOTO AQUATICS

SCOONES/SEAPHOT

Rays are flattened as an adaptation for life on the bottom. The tail is no longer used for swimming and in this sting-ray (*Taeniura lymna*) it carries the defensive sting. Rays swim using undulations which pass down the enlarged pectoral fins.

Remoras have their dorsal fins modified into efficient suckers. They live in close association with a large, fast-swimming fish, sharing its food and cleaning off external parasites. These specimens of *Echeneus neucrates* are hanging on near the gill slits of a large shark.

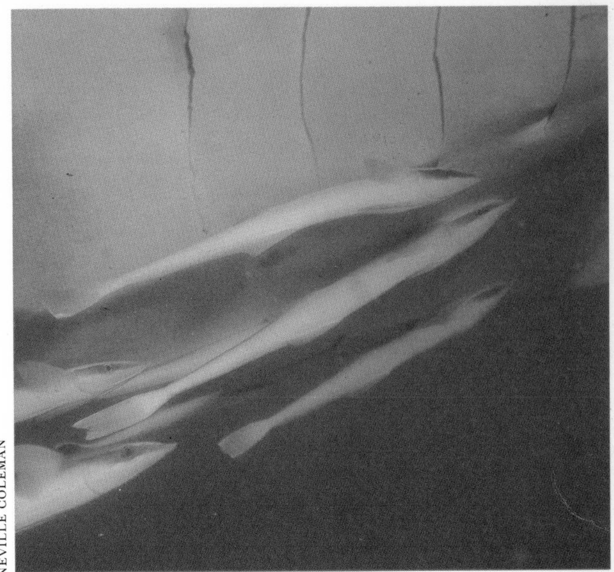

NEVILLE COLEMAN

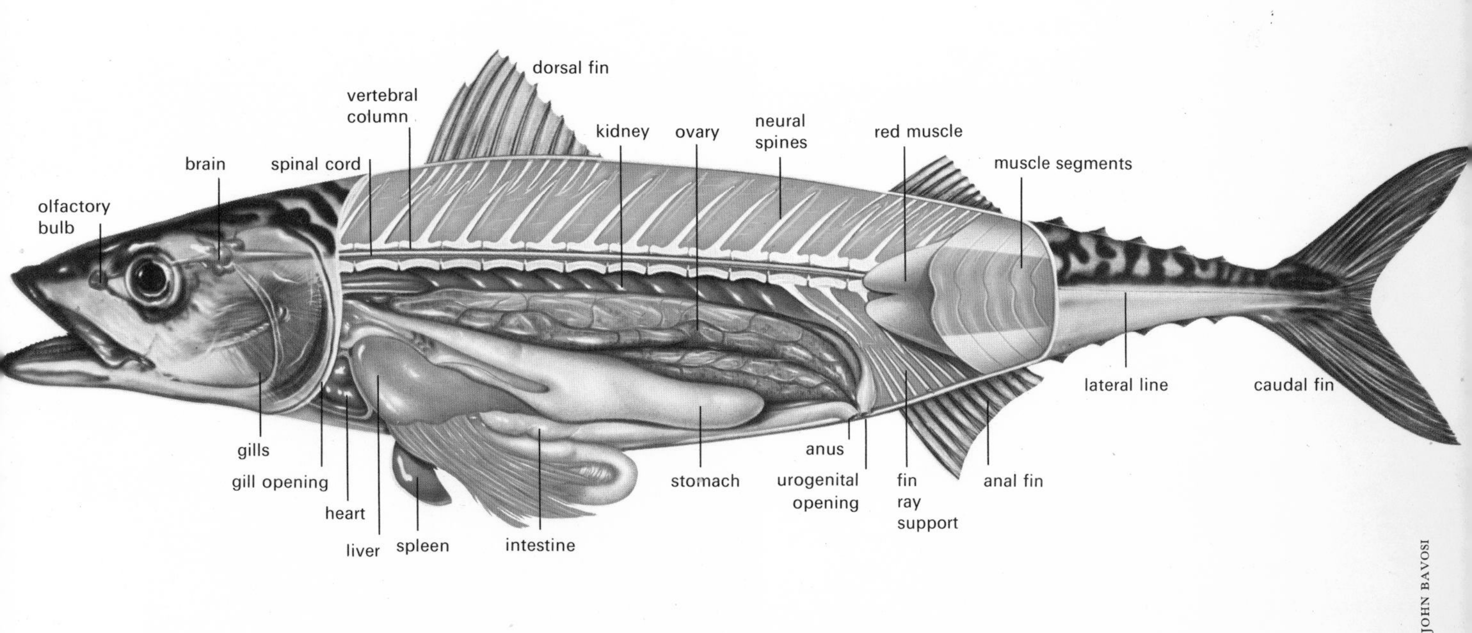


JOHN BAVOSI

Left: The brilliant colouration of the lion- or turkey-fish (*Pterois volitans*) advertises its dangerous nature. The long spines of the fins are lined with poison glands and can cause exceedingly painful stings. All warning colouration is easy to see underwater and in this case it is reinforced by the fish's behaviour.

A drawing of a mackerel showing many of the typical features of a fish's anatomy. Unlike many oceanic fishes, the mackerel does not have a swimbladder along the top of its body cavity. The conspicuous red muscle shows that it is a powerful, active swimmer, keeping buoyant by constant movement.

## Gills and respiration

Like all other animals, fish need an oxygen supply for respiration. Unless they live very close to the surface, their only possible source of oxygen is that dissolved in the water. More oxygen will dissolve in cold water than in warm water but, even so, the maximum is only 5–6 ml per litre. Fish can take up a little oxygen across their skin but this is insufficient to keep even their basic body functions working so most of the vital oxygen is extracted from water flowing across their gills. Continuously fast swimming fishes, like mackerel and tuna, maintain a sufficient flow over their gills by just swimming with their mouths open. Less active fish, like the cod, need to pump water through the gills, which they do by gulping movements of the mouth and the action of the flexible trailing edge of the gill flaps.

## Swimming

The basic mode of swimming, typical of the clupeid fishes such as herrings, anchovies and sardines, is by beating the tail, the body bending about two-thirds of the way back, so that the head end points steadily straight ahead. The propulsive force is provided mainly by the caudal fin with the pectoral fins trimming its pitch (i.e. vertical movement) and the dorsal fin controlling the yaw (i.e. movement from side to side). Clupeids have relatively flexible bodies and turning is effected by a combination of arching the body and steering with the pectorals. The scombrids—mackerel, tuna and skipjack—have much more rigid bodies. The tail beat is much stiffer and much more powerful. The rigidity of the backbone stores elastic energy as the tail arches, releasing it as it straightens on the effective stroke, reinforcing the muscular contractions. Swimming speeds vary greatly, the fastest known swimmer being the sailback, which reaches speeds equivalent to over 100 k.p.h.

Swimming in eels is serpent-like; almost the whole of the body is thrown into undulations which pass down the body from head to tail. The push through the water comes from the sides of the body and not from the tail fin. This type of movement is also suitable for slipping between stones on the bottom and between columns of coral.

In direct contrast are the fishes that have totally rigid bodies and depend entirely on the sculling movements of their fins. Box-fish are an obvious example. Their main propulsion comes from their whirling pectoral fins and the tail is used rather like

PARISH

GRUHL/PHOTO AQUATICS

Left: Dense shoals of the barber-eel (*Plotosus lineatus*) are a common sight over reefs and in estuaries of the Indo-Pacific. The outstanding banded colouration of the eel-shaped bodies is a warning that the barbed spines of the dorsal and pectoral fins are poisonous.

The spotted colour of the barramundi cod (*Chromileptes altivelis*) is an effective camouflage for this noctural fish, which by day lurks in hideaways in the reef over white coral sand. The powerful body and enormous mouth indicate that it is a greedy predator.

Left, below: The rigid bodies of box-fish are sculled through the water by their highly manoeuvrable pectoral fins, while the tail is used like a rudder. The colouration of white spots ringed with brown, on a yellow background, warns of its poisonous flesh.

Barracuda (*Sphyraena* sp.) with their heavily toothed jaws and long slender bodies, are rapacious predators and hunt in packs. The barred, bluish colouration of their flanks is a form of camouflage typical of a great variety of near surface fishes.

BARNETSON/NHPA

a rudder. Pipe-fish use their dorsal fins in much the same way, but their bodies are still flexible enough for them to have a serpentine movement when they are in amongst seaweeds or sea-fans. Seahorses use their tails prehensilely, holding their bodies vertically and sculling with their dorsal fins.

Modification in body shape has occurred in a quite different direction in the flat-fishes, which are adapted to spending a life on the bottom. In the skates and rays this flattening has occurred dorso-ventrally, whereas in the flat-fish, such as plaice, flounder, dab and soles, the flattening is lateral. A plaice larva is initially like any other fish larva, but, as it grows, one eye begins to migrate round the head. When this movement of the eye is completed, the young plaice takes up a bottom-living existence. Both eyes then lie on one side of the head which is functionally the animal's top side. The mouth does not rotate. The plaice swims either by flexing the body—in what now appears to be the vertical plane—or by passing undulations down the dorsal and anal fins on the sides of the body.

## Buoyancy mechanisms

Fish that dwell on the bottom have no problems of buoyancy, since they spend most of their time resting on the sea-bed. The scombrid fishes also need no special buoyancy mechanisms because they swim ceaselessly. In the majority of fishes, however, there is a bladder filled with gas secreted by special cells. This bladder's primary function is to maintain the fish's overall specific gravity close to that of sea-water, so that if it stops swimming it stays poised in midwater and does not sink. The swim-bladder develops as an extension of the alimentary canal and its original function was probably for breathing, as in the lungfish. In teleosts it is now used mostly as a hydrostatic organ or, in a few species, as a resonating chamber, either for sound production or sound perception. The duct connecting the swim-bladder to the gut is retained in some species but lost in others.

Sharks do not have swim-bladders so, like tuna, they have to swim constantly. In some sharks, for example the filter-feeding whale shark, the liver is very fatty or oily and this helps to keep them neutrally buoyant. In the dogfish, however, the rigid pectoral fins are permanently set at an angle so that, as it swims, the head is lifted and planed upwards. It would be impossible for the shark to swim on an even keel if the lifting force at the head end was not balanced by a counteracting lift at the

NEVILLE COLEMAN

The weedy sea dragon (*Phyllopteryx taeniolatus*) occurs amongst seaweeds on the coast of South Australia where the skin flaps and gaudy colour are an effective camouflage. Like all pipe-fish, it is a plankton feeder with a pipette-like mouth.

HEATHER ANGEL

Below: The angler fish (*Histiophryne bougainvilli*) in common with a number of shore fish, guards its eggs, which hatch into well grown larvae. Such fish lay fewer but larger and yolkier eggs than species that have planktonic larvae.

NEVILLE COLEMAN

tail end. This is provided by the shape of the tail. The top lobe of a shark's tail is much larger than the lower lobe. As the shark's tail beats, its top edge leads, and the trailing lower edge provides the necessary lift. The lift provided by pectoral fins and tail balances the weight of the shark so that it maintains its position in the water.

## Reproduction

In teleost fishes, many of the marine species have planktonic larvae, microscopic forms which develop quite independently of their parents. The females shed their eggs into the water and, at the same time, the males shed their milt. The water currents and the limited mobility of the sperm are relied on to bring the gametes together for fertilization.

Left: The flounder (*Platichthys flesus*) is a typical flat-fish. The right eye of the larval fish migrates round to its left side as it matures, and the right side becomes its 'lower' side. The side fins are modified dorsal and anal fins.

*Aulostomus chinensis* is known in Australian waters as the trumpet fish. It is cryptically coloured and often drifts at odd angles or aligned with corals and gorgonians. It slowly stalks its small, active prey which is sucked up by the long tubular mouth.

Synchronization is important for breeding to be successful and so the fish often form great spawning shoals. It is the location and exploitation of these high concentrations of spawning fish that make much commercial fishing profitable. The numbers of eggs produced by many fish are prodigious. The halibut produces over 2 million eggs each time it spawns, the cod 6½ million, the turbot 9 million, the conger eel 15 million, the ling 28 million and, highest of all, the ocean sunfish 300 million. A newly hatched sunfish larva is only about 2·5 mm long compared with its parents' 3–4 m. Needless to say the mortality of eggs and larvae is considerable, otherwise in only a few years the oceans would be choked with sunfish. The eggs laid in such massive numbers do not contain much yolk, so the larvae need to feed almost immediately. Being so small themselves they can only consume tiny planktonic animals and plants, and no matter at what depths the adults live, the larvae live in the richer subsurface layers of the ocean. This has an added advantage to the species in that the larvae are widely dispersed by currents and by diffusion. Conditions for the successful growth and survival of the larvae may occur only in limited and scattered areas which are rarely the same from year to year.

PARRISH

Many commercial species of fish have 'traditional' spawning grounds to which they migrate. The fish spawning on each ground are often identifiable as being a distinct stock, separable from other members of the same species by such things as blood groups. Each stock probably returns to the grounds on which it was originally spawned and the larvae are dispersed down current as they hatch. Plaice larvae congregate on specific nursery grounds where they settle down onto the bottom and metamorphose into their adult form. The homing instinct is particularly well developed in the salmon, which return to spawn in exactly the same stream as they spent their larval life before migrating out to sea to feed. Large shoals of maturing salmon are now being caught off Greenland. Where these fish were spawned is unknown, although some authorities believe they are European fish.

The grunnion is another example of extraordinary breeding behaviour. These little fish congregate in vast numbers at night at the top of the high spring tides along certain sandy Californian beaches. The eggs are laid in hollows in the sand and covered up. They stay buried until they are washed out by the next high spring tide, by which time they are ready to hatch.

Quite a number of shore fishes brood their eggs. In temperate seas blennies can be found beneath boulders on rocky shores, curled around their egg masses. Each egg mass contains perhaps 100–200 eggs. Since fewer eggs are produced, they contain more yolk and the larval fish are far more advanced when hatching occurs. A shore-dwelling species would not gain much advantage by having its eggs and larvae widely dispersed, since they might well be carried out to sea into deep water. There are exceptions to all these generalizations. Herring, for example, are oceanic, near-surface living fish which might be expected to produce vast numbers of free floating eggs with little yolk, like other fishes with a similar mode of life. Instead they lay their heavily yolked eggs close to the bottom and their larvae migrate upwards to the surface feeding grounds.

Brooding of eggs is carried to an extreme in several species of freshwater fishes which brood both eggs and larvae in their mouths. The sea catfish of the eastern Atlantic coast of America does the same thing, the male fasting for six weeks while he broods both the eggs and the newly-hatched larvae. Male pipe-fish and seahorses brood their eggs in a pouch (marsupium) on the body.

Sharks and rays have two common forms of reproduction. In some species the eggs are laid in special egg-cases, colloquially known as mermaid's purses. A whale-shark egg case trawled up in the Gulf of Mexico was 30 cm long and contained a developing larval whale-shark 36 cm long. Port Jackson sharks produce conical egg cases with a spiral flange. Fertilization in sharks is internal. The male fish has pelvic fins modified into claspers which are used to deposit the sperm into the female's cloaca. Whale-sharks and Port Jackson sharks are oviparous, i.e. they lay fertilized eggs. Other species are ovoviviparous—the fertilized egg, instead of being laid, is retained within the female's uterus. The young shark is born soon after the egg hatches. Throughout its development, the embryo is totally dependent on the female. This type of development occurs in the nurse-shark (*Ginglymostoma cirratum*). In the

PARISH

A male pipe fish (*Corythoichthys* sp.) carries eggs in its brood pouch, formed by a fold of soft skin on the belly. The young are retained in the pouch until they are fully developed.

Leatherjackets (*Navodon scaber*) have small mouths with sharp teeth and a poisonous spine on the front of the dorsal fin. Here they are feeding near the surface on a moon jellyfish *Aurelia*.

HEATHER ANGEL

porbeagle (*Lamna nasus*), once the embryo has completely absorbed all its yolk, it begins to feed on unfertilized eggs which pass into the uterus.

Full viviparity, with the developing embryos attached by a placenta to the mother, occurs in several of the reputed man-eaters such as the white tip (*Carcharhinus longimanus*), the great white shark (*Carcharodon carcharias*) and the hammerhead sharks (*Sphyrna*). Up to 82 embryos have been found in the tiger-shark (*Galeocerdo cuvieri*).

## Modes of life

Although the sea as an environment is far less subdivided into a mosaic of microhabitats than are terrestrial habitats, there are two distinct modes of life: the midwater or *pelagic* mode, and the bottom or *demersal* mode. In midwater, the fish communities are stratified vertically with depth (see Chapter 6). Demersal fishes can be broadly divided into two types: those that feed on animals they encounter above the bottom and those that feed on animals from off or in the bottom. Morphologically, they are often readily distinguishable by the position of the mouth. Those with mouths on the front of the head feed above the bottom; those in which the mouth is underslung feed on the bottom. The way in which the animals sense their food is also related to their mode of life. Pelagic fishes in the surface layers often feed visually. Demersal fishes use taste and smell much more and many possess barbels or other special sensory adaptations.

In shallow water, around reefs and on the shore itself, the environment is much more complex and so are the fish communities; here some species may develop mutually advantageous relationships with other types of animal. The problems of survival on a rocky coast are quite different from those on a sandy shore. Exposure to waves and currents will vary from sheltered bays and estuaries to coasts fully exposed to the pounding of an oceanic swell. Perhaps the most extreme mode of life is shown by the fish that live in the freezing waters beneath the ice of the Antarctic. Some have a type of antifreeze within their blood to stop them freezing solid; others are supercooled and will instantly freeze solid if they come in contact with ice crystals.

## Feeding adaptations

Fish feed on a whole spectrum of types of food. Relatively few adults feed directly on the microscopic plants in the water, but the anchovetta of the Peru Current actually filters them onto its fine gill rakers. Grey mullet feed on fine plant and organic debris from muddy substrates. Parrot and surgeon fishes rasp off algae growing on coral skeletons on reefs. However, the vast numbers of fish larvae that occur in the sunlit surface waters of the oceans are for the most part dependent on microscopic organisms—including plant cells—for their food. As the larvae grow, so their diets change to the animal or zooplankton. The evidence suggests that fish continue to grow throughout their life, provided that they have an adequate food supply. Some of the largest fishes are plankton-feeders. In the tropics, where there is a continuous supply of plankton, the whale-shark and the manta ray are both planktivorous. In temperate waters, the basking shark exploits the high plankton concentrations in spring and autumn, and sheds its gill rakers during the winter when little plankton is available. At the other end of the size scale are specialized feeders, such as the seahorses and snipefish, which have tubular mouths designed like pipettes, to suck up individual plankters.

Between these extremes is a hierarchy of carnivores which attack larger animals as their own size increases. The ecosystem is arranged into a food chain or, more accurately, an anastomosing food web. The primary producers are the plants—mostly the microscopic algae or phytoplankton. These are grazed upon by herbivorous animals, mainly animal plankton. The herbivores are the food of planktivorous fish, carnivorous plankton and other larger carnivores—prawns and squid. The primary carnivores are, in turn, eaten by larger and more powerful carnivores. At the end of the food chain are the giant fishes—swordfish, bluefin tuna and marlin, the big predacious sharks—the great white shark, tiger-shark and hammerheads—and predatory whales—the killer whale and the specialist sperm whale. Recycling of the organic content of the waste material at all levels in the food web, is performed partly by bacterial decomposition and also through scavengers. The same animal may double up on ecological roles—for example a shark may switch from being a vicious predator to cleaning up the corpse of some other large dead creature. Similarly the anchovetta, which for most of the time grazes directly on the phytoplankton in the rich upwelled water of the Peru Current, will switch to eating planktonic Crustacea if plants are in short supply. Fish occupy an important part of the oceanic ecosystem and it is through the fish, for the most part, that Man seeks to exploit the oceans.

The receding tide has uncovered the massive rocky foundations of this shore. At the upper tidal level, the littoral and terrestrial habitats merge in the supralittoral zone, and on the rocks can be seen the lichens and algae which characterize this type of shore.

# At the Edge of the Ocean

The seashore is that area of land which is bounded at its upper margin by the extreme high water mark of spring tides (EHWST) and at its lower by the extreme low water mark of spring tides (ELWST). Despite these obviously natural limits, modern ecologists often extend them to include the area immediately above EHWST—the supralittoral—and the area immediately below ELWST—the sublittoral zone. In fact, since its biology is, in effect, a part of the same extensive but graded ecosystem, the littoral is often considered to extend to the edge of the continental shelf at 200 m depth or more.

The supralittoral, which on gently shelving, moderately sheltered shores may be very narrow, and on steeply shelving, exposed shores very broad, is biologically interesting since, being alternately exposed to salt spray and terrestrial weather conditions, it is a sort of buffer zone between the two environments. Similarly, the sublittoral is, in many respects, intermediate between the tidal region and the deep water areas.

The littoral is generally defined as the area of seashore which is covered and uncovered by the tides twice a day. So animals which are subjected to this regime must be able to tolerate, for varying and alternating periods, cover by sea-water and exposure to sunlight, wind and freshwater in the form of rain.

The shore is formed by the interaction of land and sea, aided by the weather. Its nature will depend largely on geological factors, such as the type of rock, whether it is hard or soft, easily weathered or resistant, and on meteorological factors, such as wind velocity, which affects the wave force; rain, which washes down particles and wears away rocks; and the tidal range. Thus shores may be particulate, i.e. composed of small, more or less mobile particles, ranging in size from fine muds to sands to fist sized pebbles, or non-particulate. The latter are composed of rocks, boulders and cliffs and are characterized by the immobility of the substratum. It is the variable combinations of substratum, exposure, and geographical factors, such as climate, that dictate the nature of the shore communities.

## Who's who on the seashore

The seashore is the only marine habitat which may be studied easily and comparatively thoroughly without costly equipment. It is, therefore, the habitat best known to professional and amateur marine biologists alike and, because of its wide diversity of ecological niches and abundance of species, it has been instrumental in introducing generations of students of natural history to the study of ecology.

To facilitate identification, biologists classify animals into groups (phyla), members of which all have certain anatomical features in common and a similar pattern of cellular organization. Most of these groups have representatives among the shore creatures and they show an amazing variety of adaptations to life on the different types of shore. Because of their size and numbers, some groups will be far more familiar than others. Indeed, some may be known only to the professional marine biologist.

**Protozoa** This group is composed of single-celled or, more correctly, acellular organisms. The Foraminifera are the most abundant protozoans on the shore but, because of their microscopic size, are not immediately obvious. They secrete a shell-like test around themselves and live among detritus and sand grains. They may be important in the coral reef ecosystem as the tests become incorporated into the material which binds the reef together.

**Sponges** The organization of the sponges (Porifera) is only a little more complex than that of the Protozoa. They are multicellular but each cell functions largely as an individual. The simple sponges, e.g. the purse sponge (*Grantia compressa*), which can be found beneath rocky overhangs, are shaped like flattened vases with perforated walls. There are flagellated cells lining the interior which draw feeding and respiratory currents through the holes in the walls. Waste and reproductive products are expelled through the hole at the top. The compound sponges conform to the same basic plan but the single central chamber has become convoluted and highly branched. These are the encrusting sponges, e.g. the breadcrumb sponge (*Halichondria panicea*), and some species may be quite massive. The body of the sponge contains spicules, which act as a skeletal support. They may be made of calcium carbonate, as in *Grantia*, or of silica (Venus' flower basket), or of horny fibres (bath sponge). Spicules may also be protective. In *Sycon coronatum*, for example, they protrude through the body wall. Other sponges develop a tough outer layer for protection.

**Coelenterates** The jellyfishes, sea-anemones, corals, sea-pens and plant-like hydroids are included in this group. A great variety of form exists, particularly among the colonial animals, but the individuals all conform to a basic shape, that

HEATHER ANGEL

HEATHER ANGEL

The simple purse sponge (*Grantia compressa*) attaches itself to weeds such as *Plumaria*, under rocky overhangs. A water current which enters through small holes in the body wall leaves via the conspicuous aperture at its apex. The jointed weed is *Lomentaria*, one of the many red algae found on rocky shores.

Left, above: Even at high water the rocks, covered with yellow lichen and olive-brown *Pelvetia*, are not inundated. This is the splash zone: organisms which live here must be able to withstand considerable exposure to the air.

HEATHER ANGEL

Left: Sea anemones extend their long tentacles, armed with batteries of stinging cells, into the surrounding water. Stunned and entangled prey are then brought back to the mouth which, as in this *Actinothoe*, is an elongated slit. The red organisms are tunicates, *Distoma*.

GRUHL/PHOTO AQUATICS

Left, below: Many sponges, such as this *Aplyfina*, are massive and may cover large areas. In these compound species, the basic shape of the simple sponges is modified by folding and repetition of chambers to increase the internal area for feeding and gaseous interchange. The exhalant openings can be seen at the tip of each upright mass.

of a vase with a circle of tentacles around the top. This shape is radially symmetrical. The cells are arranged in two layers, separated by a stiff jelly-like substance containing a simple network of nerve cells.

Coelenterates are predatory. They have a unique type of cell—the nematocyst—which contains, under pressure, coiled up penetrating and entangling threads. These cells are found on the body and, particularly, on the tentacles. By a complex mechanism, in response to chemical and pressure stimuli, these cells will discharge their threads at prey, which thus become trapped and paralyzed.

Among the sea-anemones, the common snakelocks, dahlia and beadlet anemones are to be found attached to rocks, in crevices, pools or below overhangs. On sandy shores, where the substrate makes attachment difficult, some anemones have adopted a burrowing habit. Further down the shore the fascinating stalked jellyfishes, resembling upturned bells, may be found attached to stones and seaweeds.

Outer coverings or skeletal supports are secreted in the hydroids, which are found growing on large brown algae and among the corals. The corals themselves may be solitary, ranging in size from the Devonshire cup-coral, which is about 1 cm across, to the tropical *Fungia*, which may be over 1 m across, or colonial, e.g. the tropical reef-building coral, *Platygyra lamellina*.

Sea-pens, which are colonial, may be found on the lower tidal parts of tropical sandy shores and in the sublittoral of temperate shores. They have an elongated muscular burrowing region, which maintains a hold within the substratum while the rest of the colony is held more or less erect by water pressure and the support of spicules in the tissues.

**Flatworms** These are the simplest worms and, although they may be quite abundant in some marine habitats, they are often overlooked because they are generally rather drab in colour, rather small and tend to live under stones and weeds. They are bilaterally symmetrical and move in a characteristic flowing manner. At the head end there are quite well developed sense organs but the gut is very simple and food is ingested and waste expelled through a single opening. *Procerodes ulvae* is a small drab species common on the temperate littoral. *Prostheceraeus vittatus*, however, which is often abundant sublittorally, may be more than 2·5 cm long and is a rich cream colour with dark stripes along its body.

**Ribbon worms** These worms (Nemertini) are smooth, slimy and unsegmented and the gut has a separate mouth and anus. They are all carnivorous and are most remarkable for their powers of expansion and contraction. The bootlace worm (*Lineus longissimus*) of sandy shores may be 10 m long when moderately contracted yet be capable of expanding to more than twice this length. There is a record of a specimen which was 20 m long when contracted!

**Roundworms** As their common name suggests these worms, the Nematoda, are round in cross-section. They are also pointed at both ends and move with a characteristic jerky action. Numerically they are probably the most important in the marine environment. As parasites they are found in the body cavities and organs of both vertebrates

HEATHER ANGEL

Hydroid coelenterates often have an outer supporting and protecting cover. In *Tubularia indivisa* this is horny and does not cover the polyp head, allowing the tentacles which surround the central mouth to move freely.

Unsegmented nemertine worms such as *Lineus longissimus* live in mud or muddy sand, from which they extract their food. They are often immensely long but are capable of great muscular contraction. Some can expand to twice their contracted length.

HEATHER ANGEL

HEATHER ANGEL

Unlike many temperate marine flatworms, *Prostheceraeus vittatus* is large and colourful. It has quite well developed sense organs, but its gut is simple, with only one opening.

and invertebrates, and free-living species are a major constituent of the fauna of particulate shores. However, they are never large or conspicuous and identification is extremely difficult.

**Annelida** Characteristically these worms have segmented bodies bearing numerous bristles (setae). There are three major groups of annelids: the earthworms, which have only a few exclusively marine species; the leeches, which parasitize ocean living fishes and, therefore, do not properly come into a survey of shore species; and the bristleworms or Polychaeta.

The bristleworms are the most numerous and diversely specialized of worms and occupy a variety of habitats. They are divided by life style into free-living and sedentary species. The former have well developed sense organs, grasping jaws, and paddle-like projections (parapodia), arranged serially along each side of the body, by means of which they swim or crawl. They are mainly active predators and include the scaleworms and ragworms. The sedentary species either live in burrows or secrete tubes, which may be composed of calcium carbonate (e.g. *Pomatoceros triqueter*), sand mixed with mucus (e.g. *Lagis*), or mud mixed with mucus (e.g. *Sabella*, the peacock worms).

*Polymnia*, a polychaete worm, lives in a mucus-lined burrow in coarse sand or under solid objects such as dead mollusc shells—here *Pecten*. The short, pink tentacles are respiratory gills while the long, white ones are used for trapping and carrying detrital food particles.

HEATHER ANGEL

Burrowing worms feed either by ingesting organic matter in the sand, as in the lugworm, or by drawing food-bearing currents of water through the burrow, as in *Chaetopterus*. Tube-worms feed by means of tentacles which are extended from the top of the tube. The tentacles either search the surface of the substratum for food particles (the terebellid worms) or they may have cilia which create currents and draw suspended food particles towards the mouth, as in the peacock worms.

**Arthropoda** This is the largest phylum in the animal kingdom and its members are characterized by the possession of a hard, jointed exoskeleton, which is generally divided into cephalothorax (head-chest) and abdomen. The limbs along the body are variously modified for different functions—sensory purposes and feeding on the head, walking and swimming on the cephalothorax, brooding and copulation on the abdomen.

Most of its marine representatives belong to the class Crustacea, which includes the planktonic ostracods and copepods, the primitive *Hutchinsoniella,* which lives in the spaces among sand grains, and the familiar crabs, lobsters, shrimps and prawns, whose limbs and bodies are variously adapted for walking, swimming and burrowing.

HEATHER ANGEL

Bivalved molluscs, such as the edible mussel (*Mytilus edulis*), typically have a pair of tubes or siphons through which they draw in and expel a water-current. The mussel's inhalant siphon is fringed with sensory tentacles to sample the quality of the water in the incoming current.

Most marine arthropods are Crustacea, such as shrimps, prawns, lobsters and crabs. This crawfish (*Panulirus*) shows clearly the characteristic heavy exoskeleton covering the whole animal, the jointed limbs and long, sensory antennae.

NEVILLE COLEMAN

Despite their spider-like appearance and common name, the sea-spiders or Pycnogonida are not related to land spiders. This deep-water species is feeding on a nemertine, into which it has thrust its proboscis. A sea-spider's body is so slender that part of the gut extends into the legs.

The undersurface of this littoral boulder shows the wealth of animals that are often to be found in this habitat. Most conspicuous are the amber-coloured sea-slugs (*Rostangia rubicunda*), the pale-blue chiton (*Amaurochiton glauca*), the large starfish (*Coscinasterias alamarea*) and the small cushion-star (*Patiriella regularis*). There are also sponges, a nemertine, serpulid tube-worms and barnacles, as well as coralline algae.

The king 'crabs' of tropical shores are, in fact, more closely related to the spiders. They are unique in the marine world for their method of respiration, which is via a series of gill-books—thin vanes of tissue richly supplied with blood—on the underside of the abdomen. The long spine on the end of the abdomen serves to right the animal if it overturns when walking.

The Arachnida is the group of arthropods to which the spiders, scorpions, ticks and mites belong. Apart from a pseudoscorpion, *Neobisium maritimum*, which occurs on temperate rocky shores, and some mites, there are few members of this group on the seashore. Often mistaken for, but not related to, the spiders are the Pycnogonida. These have such slender bodies that the gut extends into their disproportionately long legs. They feed upon hydroids and sea-anemones and are common on the middle and lower shores.

Like the arachnids, insects are not well represented on the shore, although members of the genus *Anurida* occur in a wide variety of habitats, from temperate rocky crevices to tropical mangrove mud. The common *Anurida maritima* can be seen floating in small grey-blue clusters on the surface of rock pools. Bristle-tails are often abundant in the drier parts of the upper shore and some flies spend their larval stages in rotting seaweed. The beetle *Aepopsis rabini* spends its entire life on the shore.

The dorso-ventrally compressed isopods, which resemble woodlice (e.g. the sea slater), and the laterally compressed amphipods or sandhoppers are common on a variety of shores.

**Mollusca** This is a very extensive group with a great number of marine representatives, showing a wide diversity of form and way of life. The most primitive molluscs to be encountered on the seashore are the chitons or coat-of-mail shells. These have a series of articulating dorsal shell plates and cling to rocks by means of a broad single foot.

By far the most numerous are the gastropod molluscs—the snails, slugs and limpets. They are characterized by the possession of a single calcareous shell, which, however, has been lost in some species. Compared with their terrestrial cousins, the sea-slugs are very colourful and ornate, often with feathery gills or other processes on the back and sides. The shells of sea-snails may also be beautifully coloured, e.g. the cone shells and painted top shells, and their range of shape is considerable, from the simple cone of the limpet, to the perforated ear shape of the ormer and the spirally coiled *Murex* with its elaborate spiny projections. In some species, e.g. *Lamellaria*, the shell may be reduced and completely embedded in a fleshy mantle. The tusk shells also possess a single calcareous shell, but it is a simple cone, open at both ends. They live in small chambered burrows in sand.

In the bivalve molluscs, the animal is laterally flattened and enclosed by two curved shells (valves). There are two siphons which protrude through a gap between the valves. Through these, food and respiratory currents are drawn by cilia on the large gills within the mantle cavity. The siphons are quite short in the species which live on or near the surface, e.g. mussels, but in the deep-burrowing species of sandy and muddy shores, e.g. *Scrobicularia*, they may be of considerable length. Bivalves move by means of a muscular foot, which in the razor shell is adapted for burrowing, in the scallop for leaping, and in the piddock for boring into rock. Some bivalves can swim actively by flapping their shells, e.g. the queen scallop; others do so by means of undulating outgrowths on the edge of the mantle, e.g. *Lima*. Bivalves such as the wood-boring shipworm and the tropical watering pot, have a much modified shell, which is supported by a secondary calcareous tube.

The octopuses, squids and cuttlefish form the group of molluscs known as the Cephalopoda (head-footed). Apart from *Octopus* and *Eledone*, which often abound on the littoral parts of coral reefs, and the little cuttlefish, *Sepiola*, which may be found at extreme low water on temperate shores, they are more typical of open waters.

**Echinoderms** These 'spiny-skinned' animals, despite their variety of shape, all conform to a radially symmetrical plan in which the parts are typically arranged in fives or multiples thereof. They move and/or feed with their tube feet, which are hydraulically controlled. These feet can be seen easily on the underside of starfishes, which also use them to prise open bivalves so that they can insert their pharynx and eat the contents. In the brittle stars, whose long thin arms radiate from a small central disc, the tube feet are used solely for feeding.

In the sea-urchins, the radiating arms have been transformed into a hollow spherical test, which, in some species, such as the sand dollars, is considerably flattened. The spines on the test vary in shape from the needle sharp spines of *Psammechinus* to the pencil-like spines of the cidaroids. Many urchins are grazers of algae and encrusting organisms and feed by means of a complex jaw apparatus called

NEVILLE COLEMAN

The streamlined and often large cephalopod molluscs have well developed 'camera' eyes and prehensile tentacles, here shown in their rest position. Most of them, like these cuttlefish, are able to change colour rapidly if danger threatens, or merely to blend with their background.

Starfish (*Asterias*) feed by straddling a bivalve, attaching their hydraulically controlled tube-feet to the shell and making the mollusc gape open by sustained muscular effort. The starfish then protrudes its pharynx and eats the meat.

SCOONES/SEAPHOT

FAIN/NSP

Unlike squids and cuttlefish, *Octopus* is a truly benthic or bottom-living animal which spends much of its time lurking in rock crevices and amongst boulders, often in shallow water. Its internal organs are contained in the 'bag' which hangs down from the head, from which also arise the sucker-covered tentacles.

Brittle-stars such as *Ophiothrix fragilis* have slender arms which arise abruptly from a disc-like body. The skeletal spicules which strengthen the body-wall can here be seen clearly.

HEATHER ANGEL

In sea-urchins such as this *Echinus esculentus*, the body-form has become spherical. It grazes algae and detrital films, using the five hard jaws on its lower surface as it moves over the rocks.

HEATHER ANGEL

an *Aristotle's lantern.* This is made up of five parts and has a complex musculature; its action is rather like that of a drill chuck. Other urchins, e.g. the sea potato, live in burrows and feed on minute detritus particles.

The sea-cucumbers have a less rigid body wall than other echinoderms as the skeletal components are arranged differently. Some of the synaptid cucumbers look quite snake-like as they move among coral heads and reef flats. Others of this group may feel quite sticky to the touch, e.g. *Leptosynapta inhaerens*, which burrows into temperate sands and muds, ingesting sand as it burrows and utilizing the nutrients contained in it. Other sea-cucumbers have specialized tube feet for collecting suspended detritus.

In the deep sea, the feathery, star-like sea-lilies (Crinoidea) are attached to the substratum by a stalk but the shore species are stalked only during their early developmental stages. As adults they are free-swimming, although they spend much of their time clinging to rock surfaces. They are the only echinoderms in which the mouth is on the dorsal side of the body and they feed by ciliary currents generated along their out-stretched feathery arms.

**Hemichordata** These are represented on the shore by the acorn worms (Enteropneusta), e.g. *Glossobalanus.* They are very fragile creatures and produce a great deal of mucus which assists them to move through the sand in which they burrow and also to trap food particles on the acorn-like proboscis. The particles are then transferred to the mouth by ciliary action. In the sublittoral are found the pterobranch hemichordates. These are more sac-like in form and feed by catching plankton on branched arms which are well supplied with cilia.

**Chordata** This phylum includes those animals which have, at least in some stage of their development, a skeletal rod (notochord) along the back, i.e. fishes, reptiles, birds, mammals and also the primitive sea-squirts and cephalochordates.

Sea-squirts are common and often abundant. The adult is basically a stout bag with inhalant and exhalant siphons. A food-bearing current is drawn in through the inhalant siphon and through the walls of the sac-like pharynx. Here the food particles are trapped in mucus and transported to the gut. Waste and genital products are expelled through the exhalant siphon. Sea-squirts may be solitary, e.g. *Ascidia mentula,* or colonial, e.g. the star ascidian, *Botryllus schlosseri,* which forms flat jelly-like layers on rocks and weeds.

The cephalochordates are rather fish-like in appearance. The lancelet, *Branchiostoma lanceolata*, commonly known as amphioxus, is a well known member of this group. It lies on its back in gravelly sand with the head protruding. As in the sea-squirts, water and food particles are drawn into the pharynx by the movement of cilia on its walls. The food is extracted by a series of sifting devices and passed into the gut. The notochord in amphioxus runs the whole length of the body and, by its interaction with the muscle blocks, makes a jerky swimming movement possible.

Compared with open water fishes, the shore-dwelling fishes tend to be of unusual shape, e.g. the snakelike pipe-fish and the elongated rocklings. In many, e.g. the gobies and suckers, the pectoral fins are modified to form a sucker with which they can attach themselves to rocks. Shore fishes are also rather exceptional in laying a small number of quite large eggs which are usually guarded by the male until they hatch.

**Other groups** In addition to the phyla already described there are a number of smaller ones, which, because of their limited number of species, their size or their restricted or cryptic habitats, are not very conspicuous on the shore. Their relationships are often of particular interest to the biologist.

The so-called sea-mats are colonial animals which secrete horny or calcareous outer coverings for support and protection. Superficially the individuals resemble those of the colonial hydroids. The sea-mats are now known to consist of two distinct phyla: the Entoprocta, in which the mouth and anus are surrounded by a ring of ciliated tentacles; and the Ectoprocta, in which the anus opens outside the tentacles which are borne on a horseshoe-shaped structure (lophophore) surrounding the mouth. The latter have a hydraulic means of expansion and contraction which ingeniously relates to the pressure of the surrounding water. On account of their size, the ectoprocts are easily overlooked, although they are common in the littoral. The entoprocts may be encrusting, e.g. *Membranipora membranipora,* which forms lace-like growths over rocks and algal fronds, or self-supporting, in which case they often resemble seaweeds, e.g. the flat fronded hornwrack and the feathery *Bugula turbinata.*

The peanut worms (Sipunculoidea) are often the dominant members of muddy shore faunas, especially in tropical mangrove swamps. They are round-bodied creatures with a protrusible pharynx. The rounded peanut appearance is achieved by tucking

HEATHER ANGEL

HEATHER ANGEL

Sea-squirts may be solitary, like *Cionia intestinalis* (top) or colonial, like *Botryllus schlosseri* (above). Anatomically, they are a sac within a sac. In *Cionia intestinalis* the inhalant opening which connects the inner sac with the water is the larger. The lower, exhalant opening, leads from the outer sac into which open the rectum (grey tube) and genital ducts (white tube). In the colony, each one of the paler 'star-rays' is a complete animal but their exhalant chambers open into a common sewer or cloaca.

The individual zooids that make up a bryozoan or sea mat such as this *Retepora* are small, with tentacle-like arms which generate a feeding/respiratory current. The horny or, as in this case, calcareous covering may, however, be extensive.

BANNISTER/NHPA

the front part of the body into the part immediately behind it. The echiuroids are similar in appearance but lack this power of introversion. Bristles similar to those found in annelids are embedded in their body walls and some species, e.g. *Thalassema neptuni*, are quite colourful. They usually live in crevices, embedded in detritus, and so they are not easily found.

The lophophore structure found in the ectoprocts is also found in the Phoronida and Brachiopoda. The Phoronida are elongated, tube-dwelling animals, which may live by themselves in spaces in shells or, as in some tropical species, in association with the membranous outer coverings of burrowing anemones, or even in pseudocolonies of entangled tubes in the sublittoral. The brachiopods are enclosed in a pair of valves and so at first sight look very much like a bivalve mollusc. However, the relationship of the body to the valve is dorso-ventral rather than lateral as in the molluscs. There is usually a projecting muscular stalk (peduncle) which anchors the animal to the substratum. This may be short, as in *Terebratulus*, a sublittoral genus with hinged valves, closely resembling a clam or cockle, or long, as in *Lingula*, a littoral burrowing species which uses the peduncle to anchor itself to the bottom of its burrow.

The arrow worms (Chaetognatha) are mainly planktonic but one genus, *Spadella*, is found on the shore. They resemble their planktonic cousins but are modified for existence among the holdfasts of algae by having a special attachment organ at the posterior end of the body.

Such is the variety of animals which, by careful searching, may be found on the seashore. In considering the nature of the different types of shore and the problems which they present to their inhabitants, we shall better appreciate the adaptations exhibited by the shore animals.

## Particulate shores

Particulate shores owe their ecological character to the size of the particles—which in turn governs the size of the spaces (interstices) between them—and to the degree of instability of these particles under the influence of a covering tide. Very fine (colloidal) particulate shores have only very small interstices, retain a great deal of water, which is uniformly distributed when the tide ebbs, and are moderately stable when the tide flows. Shores of moderately sized particles—sands—have moderately sized interstices and retain water in a graded manner, the greatest amount accumulating at LWST. They are more or less unstable—at least on the surface—under tidal influence. Coarse particle shores (shingles and pebbles) have large interstices, drain almost completely at ebb tide and are unstable during tidal flow.

Despite their instability, particulate shores are by no means devoid of plant cover. Tropical green algae, such as *Avrainvillea*, have a large triangular base (holdfast) composed of fibres which are able to bind into the sand. They provide little cover, however, since their fronds are small and the individual plants are scattered. A certain amount of cover is provided by plants such as the temperate *Zostera* and tropical *Enhalus*, which may form large beds near low water and whose extensive rooting systems serve to bind together and stabilize the mobile substratum.

Where there is no general plant cover, animals of these shores live within the substratum, actively burrowing through it, making temporary or permanent tubes or occupying the interstices. On the highly mobile shingle and pebble beaches, only a few crustaceans can survive between the stones.

Few large algae are found on fine particle shores but the surface layers, especially where there is mud to conserve water during ebb tides, can support a very rich meadow of diatoms. On tropical shores, where productivity is high, this cover may be so dense as to give the shore a distinctly brown appearance. It is these single-celled plants that provide the photosynthetic primary production upon which the ecological economy of much of the shore's varied groups of animals depend for survival.

## Muddy shores

Marine muds are almost invariably of terrestrial origin. The suspended particles, carried down in rivers, clump together (flocculate) when the river and sea waters meet and settle to form estuarine mud banks. Of all particulate shores, the estuarine muds show the greatest amount of ecological variation according to geographical position. In the tropics, it is in such situations that mangrove swamps, forming a major separate ecosystem, become established (Chapter 4).

It is difficult to say whether the number of animal species of a mud bank is restricted by the influence of fresh and salt waters or by the special characteristics of the mud itself. The ragworm, *Neanthes diversicolor*, is physiologically restricted to dilute sea-water, although it can tolerate a moderate range of dilutions. It has also evolved the ability to coat its muddy burrow with mucus, thus forming a membranous tube which protrudes through the burrow entrance and acts as a food-filtering device.

The surface film of diatoms and settled detritus is used as a source of food by mud-dwelling species in two ways. It may be grazed at the surface by gastropods (e.g. the temperate *Hydrobia* or the tropical *Cerithium*), or it may be sucked off the surface by animals firmly buried in the more solid mud below. Such a species is *Scrobicularia plana*, which lives buried in clayey muds at depths up to 20 cm. Its siphons, which may be up to 28 cm long, reach up through a channel to suck in the deposits of detritus and diatoms which are within reach of their mobile tips. It is thus physically protected (although some wading birds can reach it, even at this depth), while taking advantage of the rich surface food supply.

Few species are able to burrow freely in mud, although some, e.g. the amphipod *Corophium*, can excavate pits and short burrows in the sloppier surface mud. Even the relatively streamlined worms tend to restrict their burrow systems to the soft mud just below the surface. Only the more massive, muscular and streamlined worms, such as the free-living *Nereis virens*, venture further afield.

Although they do not tunnel extensively, the burrowing shrimps, e.g. *Callianassa* and *Upogebia*, can excavate and maintain considerable burrows in the mud. Most of these shrimps share their burrow, or at least the outer part of it, with other species. Some very complex associations are found on tropical muddy shores, some of them developing into separate ecosystems in their own right.

A tropical particulate shore often has numerous shallow depressions on its surface, which hold pools of water at low tide. From these come the loud clicking sounds of the pistol prawns (*Synalpheus*) as they trigger off their modified pincer to discourage potential marauders. Closer inspection reveals that the prawns are sharing the pools with small fishes, usually gobies, which take advantage of the prawn's excavating ability and of the fact that the depressions outside the prawns' burrows remain flooded at all times.

The interstices of muddy shores are so small that few animals can occupy them. Nematodes and bacteria populate this habitat extensively and, in turn, provide some of the other muddy shore inhabitants with a food source. If too much organic material is trapped in the mud, the bacteria, using

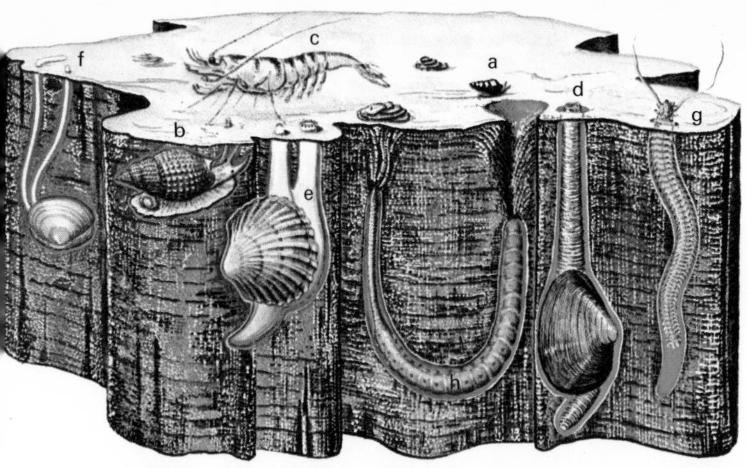


Modern ecologists think of marine animals as communities with common ecological requirements rather than as miscellaneous collections of individuals. The study of benthic communities was pioneered by Gunnar Thorson who categorized the assemblages according to the most conspicuous or dominant species. The *Macoma* community is characteristically composed of the surface-dwelling *Hydrobia* (a), *Nassarius* (b) and *Crangon* (c); the burrowing bivalves *Mya* (d), *Cardium* (e) and *Macoma* (f) and the worms *Pygospio* (g) and *Arenicola* (h). All the species are dependent upon the surface layers for their food except *Nassarius,* which preys upon the buried molluscs.

HEATHER ANGEL

Mussels (*Mytilus edulis*) may form extensive beds, attaching themselves to one another with strong byssus threads. When a bed like this forms in an estuary, the water-borne mud which settles on it forms the basis for the establishment of a specialized ecosystem where other animals can survive.

it as a source of energy, may produce poisonous by-products which prevent other animals from living at the same level in the substratum.

Despite its intrinsically mobile nature, mud can occasionally support fixed populations. Large reefs of mussels (*Mytilus edulis*) may be built up in estuaries, largely because they can attach themselves firmly to solid objects—including other mussels—by means of their tough byssus threads. They thrive in such situations, filtering the abundant plankton and suspended organic solids from the estuarine water.

These reefs are seldom, if ever, composed simply of mussels. Once established, the spaces between the mussels silt up and provide habitats for species more usually associated with sandy shores, e.g. the sand mason worm, *Lanice conchilega*. A mussel reef also supports species associated with a solid, rather than a particulate habitat, such as the slipper limpet, *Crepidula fornicata*, which is also a pest of oyster beds in temperate estuaries, and the periwinkle, *Littorina saxatilis*, which grazes on accumulated diatoms and detritus. These solid areas of shell also provide one of the few habitats for barnacles in muddy estuaries. Complex biological associations often occur. The little pea crab, *Pinnotheres pisum*, lives in the mantle cavity of the mussels, cleaning debris from the gills in return for shelter and food, and apparently entering and leaving the mussel whenever it wants to.

## Sandy shores

The clearly defined and biologically unusual supralittoral zone often found on sandy shores is one of the more interesting aspects of their ecology. At their landward edges, areas of accumulated sand often build up and, over long periods of time, form extensive sand dunes. In temperate climates these are populated largely by terrestrial species, e.g. sand wasps and tiger beetles, which usually live in sandy heathlands and probably flourish in the dunes from lack of competition.

On tropical sandy shores, the supralittoral fauna is seen at its most extensive and interesting. On the seaward edge of the zone are found the ghost crabs (*Ocypode*), so-called because, being sand-coloured, they seem to appear and disappear against the sandy background. As they occur at the junction of the tidal and supratidal zones, they display a fascinating split biological personality. For most of the day they live in burrows, which extend through

NEVILLE COLEMAN

HUTSON/NSP

Ghost crabs such as *Ocypode cordimana* make deep burrows near the high-water mark of tropical sandy shores. At twilight the crabs emerge to forage for food amongst the surface layers and at the water's edge.

Ripple-marks formed in the surface of sand-flats by tidal action can be important in the sandy shore ecosystem. On tropical shores extensive diatom populations may occur here and these, in turn, will provide food for grazing animals.

the dry sand to the cooler, moister layer below. In the evening, they come out to forage near the water's edge for microscopic plants and detritus and, occasionally, for larger pieces of vegetation.

Further inland, in the dry dune areas around the bases of plants such as the euphorbias, are the land hermit crabs (*Coenobita*), which are perfectly adapted to the demands of their environment. Using a specially modified pincer, they can seal off the opening of the 'borrowed' mollusc shell in which they protect their unarmoured abdomen, and into which they can withdraw, almost as well as the original inhabitant did with its operculum. They also carry about with them a small quantity of sea-water in the gill chamber for respiration—returning to the sea occasionally to replenish it. In spite of these modifications, they have retained normal crab reproductive behaviour and planktonic larval development and so must return to the water to breed.

This line of development has lead to one of the most curious of all marine ecological phenomena—the life cycle of the robber crab, *Birgus latro*. This very large hermit crab lives beyond what would normally be regarded as the supralittoral. It feeds upon coconuts, which it obtains by climbing coconut palms, using specially modified limbs. The coconut plantations may be some distance from the sea, but, like *Coenobita*, *Birgus* is entirely dependent on sea-water for respiration and reproduction. It is beautifully adapted to its habitat and its large gill chamber, which can store a correspondingly large volume of sea-water, is made more efficient by the presence of respiratory surfaces in addition to the gills.

Lower down the beach, in the tidal area of the sandy shore, animal communities are distributed in two distinct ways, according to the nature of the sand and to the way in which the animals cope with it and obtain their food. Actively burrowing species live mainly around the low water mark, where the sand retains most water and softens under pressure. The more sedentary and surface-dwelling species can live further up, where the sand drains more thoroughly and hardens under pressure. At first sight, there may be very little sign of life on sandy shores, as most of the animals live either permanently under the surface, or come out only at flood tide, burying themselves when the water recedes to protect themselves from drying out until the tide comes in again.

Like muddy shores, the surface layers of sandy

BONE

The land hermit crabs (*Coenobita*) spend most of their lives in the supralittoral of tropical sandy shores. However, they must retain contact with the water, returning to the sea to replenish the reservoir in their gill-chambers and to breed.

HUTSON/NSP

Perhaps the most curiously adapted of all crabs, the robber crab (*Birgus latro*) lives in coconut groves and feeds upon coconuts. Like *Coenobita*, it carries a reservoir of water in its gill-chambers and must return to the sea to breed.

shores support films of diatoms and detritus particles. Although sand drains more thoroughly than mud, water may be retained at the surface in either of two ways. As the tide recedes, the waves leave wave patterns or ripple marks in the sand. These trap water in which diatom films can develop and, in the tropics particularly, this habitat often supports large numbers of grazing and filter feeding gastropods. Most sandy shores contain a certain amount of mud in the surface layers. This results in greater water retention than would be the case with clean sand, promoting a better growth of diatoms.

Though there are many gastropods on sandy shores, they are not the most elaborately modified of the surface grazers there. Tropical sand flats are usually littered with pea-sized sandy pellets. These occur in the same area of shore as the numerous populations of soldier crabs (*Dotilla*), which may be seen scuttling in pink waves across the sand as the tide ebbs. The pellets are not, as might be supposed, the crabs' faecal droppings, but sand grains rejected during feeding. The feeding limbs, which in raptorial species act as pincers, in *Dotilla* are modified instead into spoon-shaped structures which shovel up sand grains, together with food particles, into the lower part of the

Soldier crabs (*Dotilla*) are so-called because of the way they move in great armies across the shore at low tide. Some of the crabs in this 'column' can be seen picking up clawsful of sand, from which they extract detrital food.

POLUNIN/NHPA

mouth. The food matter is extracted and the sand grains are passed upwards onto a sort of shelf just below the crab's eyes. When the mucus impregnated pellet has grown to an uncomfortable size, the crab interrupts its 'spooning' momentarily to sweep it off.

This pellet formation is put to good use as part of the crab's behavioural response to the incoming tide. *Dotilla* does not move about actively when the tide floods and, to escape being swept about too violently by the water, it makes a burrow for itself, remaining inside until the tide ebbs. At first the burrow is a shallow pit, which the crab enlarges and deepens by moving around the periphery, feeding as it goes and building up the discarded pellets around the pit edge. In this way an igloo is built over the crab and, when the tide comes in, the water pressure consolidates it, forming an effective plug to the burrow, which by this time has been deepened further.

Many other surface dwellers have developed efficient means of burying themselves, though they usually do so at the ebb not the flow of the tide. Starfishes use their mobile socketed spines to excavate the sands so that they sink into star-shaped depressions. The ridge of sand thrown up around the edge of the depression falls over the animal and covers it. A similar mechanism is used by the sand dollars, although they can also move about beneath the surface.

Other crustaceans have anatomical modifications for burrowing. *Macrophthalmus* is a crab with greatly elongated eye stalks which enable it to see above the surface when the rest of it is buried. Mole crabs, which sometimes occur in enormous numbers near the water's edge on tropical shores, dig themselves in with an efficient prong-like limb. They move up the shore with the incoming tide, keeping just below the surface and making the strip of sand ripple ahead of the advancing sea.

The masked crab, *Corystes cassivelaunus*, is an outstanding example of a crab highly modified for life on temperate sand flats. It has long antennae which are, in effect, two half tubes, each composed of an antennal stalk with lateral fringes of bristles. When placed together, the bristles intermesh, forming a tube which can be pushed upwards by the buried crab, to emerge above the sand surface like a snorkel, supplying the gill chamber with constantly changing water. The masked crab also has very elongated chelipeds with which it can reach up through the sand to grasp its prey.

The tropical swimming crab, *Matuta lunata*, has probably the most efficient mechanism of all for digging in. As well as the paddle-shaped fifth legs, which are common to all swimming crabs, it has a pointed paddle- or ploughshare-like tip to each walking leg. When stranded by the tide it settles onto the surface of the sand, spreads out its legs and, with a twisting action, almost screws itself into the sand in a few seconds—leaving only its eyes exposed above the surface.

The majority of species in the sandy littoral live permanently within the sand. Stationary or semi-stationary animals, such as the bivalve molluscs, lie below the surface, with only their siphons protruding. They can adjust their position relative to the surface by means of their muscular foot or, should they become stranded at the surface, they can use it to dig themselves back in. Tropical brachiopods, such as *Lingula*, live in burrows with a slit-like opening, through which they draw a feeding and respiratory current with the cilia on their tentacles.

A stationary way of life has led, in some species, to complicated behavioural and morphological modifications. Some sea-urchins, e.g. *Echinocardium* and *Spatangus*, use their trowel-like spines to excavate burrows with channels leading up to the sand surface and sideways into the sand. These serve respectively as a source of fresh sea-water, oxygen and food, and as a sewer. The sea-urchins keep the channels clear and in proper shape with highly adapted tube feet.

Tusk shells and pectinariid worms have evolved along roughly parallel lines as both are elongated animals living in tapering horn-shaped tubes, linked to the surface by a narrow channel. In an evolutionary sense it is probably not a long step from the situation where an organism lives in a tube, but keeps a certain amount of mobility, to one where an organism lives in a tube and is permanently anchored to one spot. Even in such cases, all freedom of movement may not have been forfeited since some tube-dwelling species—notably some of the polychaetes—can escape from their tubes if attacked.

The most conspicuous of the fixed species on a sandy shore are the burrowing sea-anemones, which often have colourful tentacles, and the worms with their protruding tubes. The temperate species of sea-anemone make either unlined burrows or burrows thinly lined with mucus but some tropical species, e.g. *Cerianthus*, secrete many-layered envelopes, which may form a recognizable tube and, incidentally, provide a habitat for other

HEATHER ANGEL

The masked crab (*Corystes cassivelaunus*) is well adapted to life in sandy shores. When placed together, its antennae form a tube through which water-currents can be drawn even while the crab is buried. Its exaggerated pincers allow it to reach out along the sand surface for food particles.

HEATHER ANGEL

Molluscs which plough through the sand must protect their shells from wearing away. The necklace shell (*Natica catena*) has a broad foot which is partly wrapped around the shell and partly extended to form an efficient sand-plough.

enterprising tube-dwelling or burrowing species.

Tube-dwellers use different types of material for their tubes, though most use mainly sand or mud particles held together by a mucus base. They may be tough and rubbery, or more rigid with grains cemented together rather like a mosaic pavement. The sand mason worm, *Lanice conchilega*, uses sand grains and pieces of shell gravel and the top of its tube is closed over and perforated rather like a pepper pot, so that the worm's feeding tentacles can emerge. To deceive predators, the areas between the perforations support a number of tentacle-like structures which are, in fact, made entirely of mucus and sand—a gritty and indigestible mouthful for a predator.

Another worm, *Owenia fusiformis*, is able to select sand grains critically. These are put first into a pouch-like organ. If they fit, they are then transferred to the edge of the tube, where they are arranged rather like the slates on a roof. This makes the tube flexible and, incidentally, difficult to bury 'against the grain'. Because it has to select sand grains, this species is restricted to those areas of shore where grains of the correct size occur.

Though most bivalve molluscs are stationary rather than fixed, there are exceptions. The tropical genus *Brechites* (the watering pots) is truly bivalved for only the larval part of its life cycle. In the adult, the larval shells become embedded in a much longer tube which, as far as is known, remains permanently fixed in the sand.

Free-living species occur both at the surface and in the substance of the sand. Active surface-dwellers are usually grazers, e.g. *Cerithium*, or carnivores which seek out their prey by digging for the buried species. The carnivores include starfish, such as *Astropecten*, but it is the snails, like *Natica* and the tropical *Polynices*, that are the most efficient.

*Natica* has a broad muscular foot upon which it glides beneath the surface of the sand. Drawing a shell through such an abrasive medium would normally wear it away rapidly but *Natica* protects its shell by wrapping its foot around most of it, thus providing a lubricated lozenge-shaped surface which passes more easily through the sand. *Polynices* is even more efficient; its foot is folded back over the shell, forming a muscular wedge by means of which the snail can plough its way through the sand at a surprising speed.

Other species have modified their anatomy so that, although the head, foot and part of the body dig down and through the sand, the shell stays above the surface. In some tropical and subtropical snails, the shell is ornamented with spines and other projections which would, in any case, make it impossible to draw through the sand. The snails feed on buried bivalves, rasping holes in their shells, and extracting the tissues with their toothed radula.

The majority of free-living species burrow through the sand, only coming to the surface when they are accidentally uncovered. Although they may belong to a wide range of phyla, they usually have a long worm-like body. This streamlining, coupled with powerful muscles and some form of skeleton, enables them to move actively, especially through the softer sand at low water mark. Many worms produce mucus to assist their passage, in the same way as acorn worms. In most species it probably does no more than lubricate and reduce friction, but in some it may be oozed into the surrounding sand, where it causes the grains to stick loosely together, forming a temporary tunnel.

Indiscriminate feeding is practised by several free burrowing species, e.g. burrowing sea-cucumbers, which take in sand as they burrow and digest out the micro-organisms and detritus. Other species have well-developed jaws for grasping their prey; some worms for example have a series of jaws, contained in a pharynx that may make up more than a fifth of the body length, giving a considerable reach. Strong jaws, however, do not always mean that the animal is a carnivore. In some sand worms they are used for no more sinister a purpose than to prevent slippery, partly-decayed plant material from slipping as it is ingested.

In addition to these macroscopic animals, sandy shores also support enormous populations of microscopic creatures. These—the protozoans, coelenterates, nematodes, etc.—occupy the interstices between the sand grains and probably contribute a great deal to the general economy of the shore.

Animals of the sandy shore are interdependent in many ways—as predator or prey, as burrow-sharers, reef-builders or sand-consolidators. Two, however, have evolved a truly integrated, symbiotic relationship, the flatworm, *Convoluta roscoffensis*, and a unicellular green alga which grows in its tissues. This flatworm sometimes occurs in such numbers on warm temperate shores that quite large areas of shore may appear green at low tide. Like *Dotilla, Convoluta* appears from just beneath the sand surface as soon as the tide ebbs and buries itself when the tide turns. This rhythm is constantly maintained, because, although it feeds like other flatworms when a juvenile, once it has acquired its green algae, the digestive system gradually degenerates and it becomes increasingly dependent on the food obtained from the algae. This behavioural pattern is, therefore, vital to ensure the best possible photosynthetic conditions for the algae. Nevertheless, the flatworm eventually destroys itself by taking so much food from the algae that they are removed from the tissues. However, since the green alga is commonly found living freely, it is always available to reinfect the next generation.

## Rocky shores

Compared with particulate shores, rocky shores may be regarded as stable. However, this stability is only relative; cliff faces and rocky platforms are undoubtedly stable but boulders, even large ones, may be shifted by the tremendous forces of the sea. The instability of the different types of shore have different effects. Thus fine particles tend to scour and abrade while wave-activated stones and boulders crush and grind.

The geology of a rocky shore is usually important to the ecosystems which it supports. Rocks vary in texture from soft sandstones to hard granites. Sandstones and limestones may be eroded into gaping pockets or holes, whereas slates may weather into elongated slits, often of great complexity.

The arrangements of the boulders in relation to the layer below may also be significant. If they lie on shelves or platforms of hard rock, or if the shore is composed entirely of such shelves, there may be comparatively little accumulated detritus, mud or sand, and the plant and animal communities will be quite different from those found on particulate shores. Conversely, the boulders may be embedded in accumulated materials and the plants and animals will vary accordingly. Most rocky shores fall into some grade of the latter category.

The physical and meteorological influences upon a rocky shore can be tremendously variable. Exposure to wave action, for example, ranges from almost complete shelter, as in a deeply inset bay protected by headlands and offshore reefs or islands, to the total exposure of the seaward-facing sides of unprotected rocky headlands. Between the two exists a wide range of shores, exposed to an equally wide range of conditions, several possibly influencing different areas of the same shore.

Perhaps the most conspicuous difference between particulate and non-particulate shores is the extensive algal cover to be found on the latter. The larger algae can attach themselves firmly to the rocks with their branching holdfasts and are able to withstand much of the buffeting to which they are subjected. The animals too are able to attach themselves to the rocks or to the algae and are protected from drying out by the algal fronds. The animals of rocky shores are, therefore, as a rule much easier to see.

The distribution of the algae gives the first indication of the different physical conditions prevailing at different levels of the shore. In temperate climates, the occurrence and distribution of algae can give valuable information about the degree of exposure on the shore, information which is often unobtainable in any other way. Some algae, for example, will only flourish in sheltered conditions; others grow differently in sheltered and exposed areas. Some species grow only in exposed con-

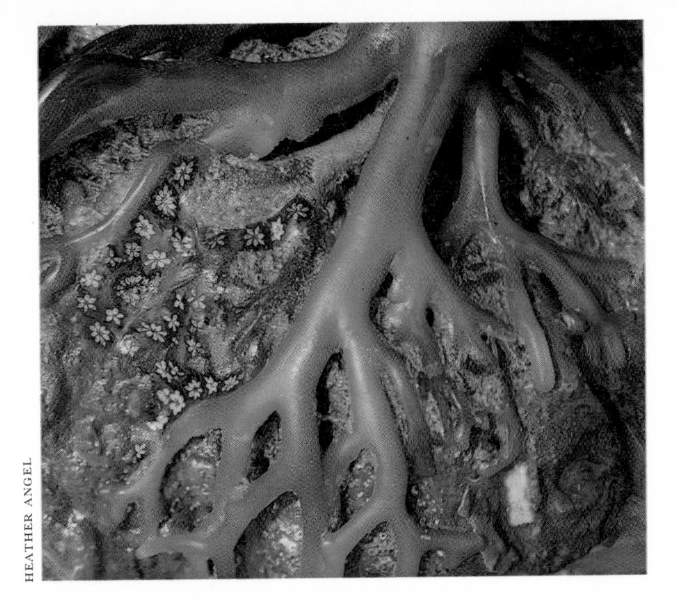

HEATHER ANGEL

HEATHER ANGEL

Because of the solid nature of the substrate, plants and animals are able to attach themselves permanently to rocky shores. Here, a fixed colonial tunicate, *Botryllus schlosseri*, shares a stone with the holdfast of *Laminaria digitata* and a pink cement-like mass of coralline alga *Lithothamnion*.

Right, above: At and just above the High Water Spring Tide level on temperate rocky shores is a narrow band of the highest 'zone weed', *Pelvetia canaliculata*. This alga is able to withstand considerable desiccation, often appearing to dry up completely in a hot summer but reviving again when wetted by the autumn high tides.

*Fucus vesiculosus*, another common 'zone weed' of temperate rocky shores, occurs below *Pelvetia* on the shore. The gas-filled bladders on the leaves keep them afloat at high tide and carry a film of micro-organisms, grazed here by yellow and dark green flat-topped periwinkles (*Littorina littoralis*).

Below: Some rocky shore weeds are so specifically adapted to environmental conditions that they may be used as reliable indicators of these conditions. Lush growths of *Ascophyllum nodosum* occur only in sheltered places and the olive-green colour of the plant makes it and the areas which it indicates conspicuous.

HEATHER ANGEL

HEATHER ANGEL

ditions, even on cliffs facing directly into the most fiercely-breaking seas.

The distribution of algae between the tide marks is important. It shows how tidal activity can influence the biological boundaries of an ecosystem and forms a foundation for several smaller ecological units within the general shore habitat.

Different species of algae can withstand different degrees of exposure to the various conditions (cover/uncover, drying out, pressure, sunlight, etc.) which are imposed on them by the movements of the tide. Under these influences, they tend to grow in distinct horizontal bands or zones across the shore. The species which can tolerate the greatest degree of exposure to non-marine conditions form the highest zone, at, or even slightly above, high tide level, while those that are least adaptable occupy the zone at, or below, the low tide mark. These zones depend on tidal activity but are also influenced by local conditions. Some brown algae, for example, are unable to attach themselves to sandstones and are therefore not found there at all. Progressing from the splash zone to the sublittoral, successive bands of brown algae will be found. The channel wrack (*Pelvetia canaliculata*) is found at the highest point on the shore, succeeded by the flat wrack (*Fucus spiralis*), bladder wrack (*Fucus vesiculosus*), knotted wrack (*Ascophyllum nodosum*), the serrated or saw wrack (*Fucus serratus*) and lastly, at the lowest point, the kelps or oarweeds (e.g. *Laminaria*). On steeply shelving shores, the zones will be narrow, while on a gently shelving shore they will be wider.

The width of the supralittoral zone of a rocky shore is related to gradient. The animal life of this zone is limited and often tends to merge with that of the highest littoral zone. It usually contains some insects and crustaceans and sometimes terrestrial species.

The fauna of the tidal part of the shore may be divided into three major ecological groupings—the splash zone (furthest from the low tide mark), the midshore and the lower shore.

The splash zone, characterized by the highest zone weed, *Pelvetia canaliculata*, has a restricted and very specialized fauna. All the species that live there, both plant and animal, are able to withstand prolonged exposure to the air or even actual dessication. *Pelvetia* is often found dried to a crispy consistency in a hot summer, only to be restored to normal after wetting by the autumn spring tides. The most conspicuous animals here are the barnacles, which often cover rocks and cliff faces so densely that the rock itself is completely obscured. Among the empty shell plates of dead barnacles may be populations of periwinkles, *Littorina neritoides* and the pulmonate *Otina ovata*. These molluscs are also well adapted to withstand dry periods. *Littorina* can seal itself into its shell with a horny operculum, secreting a thin hardening layer of mucus around the edges. *Otina* has no operculum but, when conditions become intolerable, it burrows down into the holdfasts of *Pelvetia* or of the various lichens which occur in this zone.

Splash zone lichens form the basis of an important microhabitat. Attached by a few byssus threads among the holdfasts of *Lichina pygmaea*, may be large numbers of a little pink bivalve, *Lasaea rubra*. Unlike most bivalves, *Lasaea* has separate siphons, a logical arrangement in a species which does not burrow and which is capable of some movement by gliding along on its foot. In suitable conditions, *Lasaea* may be found from the splash zone down to mid-tide level.

The fauna of the lower parts of the rocky shore is extensive—both in numbers of species and numbers of organisms. The habitat is complex, requiring many special adaptations. It is the solid nature of the shore which dictates many of the

Limpets (*Patella vulgata*) are not fixed, but move surprisingly little as they graze. When exposed by the tide they clamp their shell down onto the same area of rock with such regularity that the skirt of the shell wears an annular groove in the rock.

HEATHER ANGEL

biological reactions and interactions of the species. Here, as in all ecosystems, food and feeding play an important role, but adaptive methods of reproduction and distribution are equally important.

Rocks and boulders often have extensive diatom cover, which at certain times of the year is supplemented by juvenile algae. These meadows of primary producers are grazed extensively by several molluscs but notably and most conspicuously by the limpets (*Patella*). This common temperate animal occurs from the splash zone sometimes as far as the low water mark. In the splash zone it often occurs in large numbers on boulders which do not appear to have any kind of grazable material but, even here, a thin layer of diatoms is always present.

The limpet may be quite a large animal, up to 6 cm long. As such, it needs a regular and adequate supply of food and might therefore be expected to be far-ranging. Close examination of the area immediately surrounding a limpet will reveal a series of fan-shaped marks, radiating outwards. These are traces left by the limpet in the diatom film and show that, in fact, it only makes short grazing forays, returning each time to the same base, which is worn away by the constant touch of the skirt of the shell brushing against the rock.

Not all grazers are so restricted. Periwinkles and top shells range more freely, although they may be restricted to a particular zone.

Grazing on the rocky shore is not confined to vegetable food. On the lower shore, the thick growths of compound sponges are often grazed by sea-slugs. These handsome animals often feed on a particular type of food and are frequently camouflaged to match its general appearance. *Archidoris pseudoargus*, for example, matches the sponge, *Halichondria panicea*, in general tone and pattern, while the brick red sea-slug, *Rostangia rufescens*, may match almost exactly the red colour of sponges such as *Myxilla incrustans*. The sea-slugs probably achieve protection from predators from their colouring, which may well be due to absorption of pigments from the sponges they eat. They are some of the few animals which can feed on sponges, despite the spicules which they contain. The sea-slugs often incorporate the spicules into their own skin as they feed.

The predatory feeding habits of the cowries are similar to grazing. They rasp away at the soft tissues of sea-squirts, which often occur in sufficiently large patches to qualify as 'meadows'.

Sea-slugs, many of which, like this *Phyllidia*, are spectacularly coloured, graze upon colonial sponges. Spicules make sponges unsuitable as food for many animals, but many sea-slugs are able to incorporate them into their own body wall.

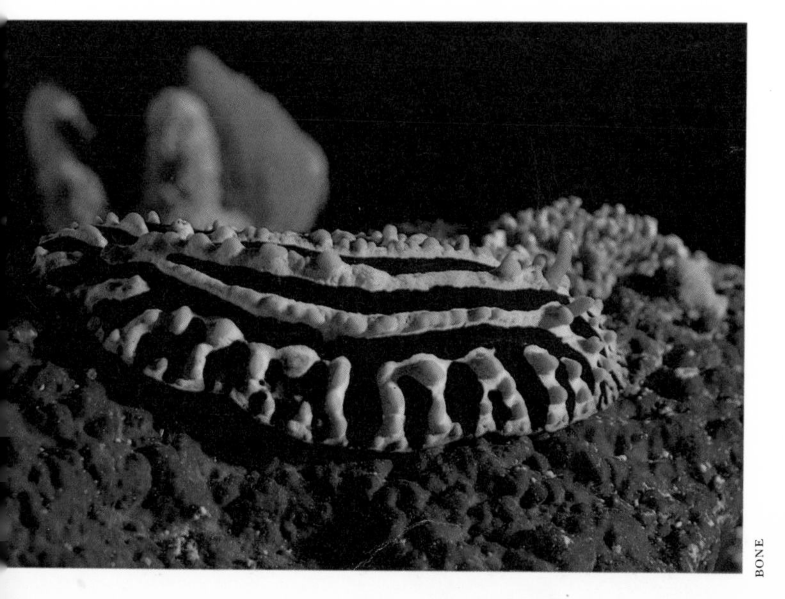

BONE

Although more abundant in the sublittoral, the whelk *Buccinum undatum* is sometimes found at Low Water Spring Tide level on rocky shores. It has a conspicuous siphon through which it draws a current of water for respiration. The operculum on its muscular foot protects the animal when it is withdrawn into its shell, while a horny covering preserves the large shell-whorl.

HEATHER ANGEL

The free-living rocky shore species have feeding mechanisms similar to those found in sandy shore species. The dog whelk, *Nucella lapillus*, the sting winkle, *Ocenebra erinacea*, and other carnivorous gastropods, rasp and etch holes in the shells of molluscs and barnacles and extract the meat with their long radulae. Crabs, e.g. the shore crab, *Carcinus maenas*, seize their food, which is usually dead or partly-decaying material, in their pincers, tear it to pieces and feed it into their shredding mouthparts. Similar raptorial feeding habits are found in other crustaceans, such as the squat lobsters and prawns. Still others, e.g. sea-slaters and amphipods, are detritus feeders.

Detritus feeders on the rocky shore include some of the echinoderms, e.g. feather stars, brittle stars, which have fascinating mechanisms for collecting food. The sea-cucumbers can move about on their tube feet but usually remain clinging to the undersides of rocks, collecting detritus particles with their outspread tentacles. Two of the tentacles have the special task of collecting up the accumulated food from the others and pushing it into the mouth.

The little crabs, *Porcellana*, are also filter- or deposit-feeders. Although equipped with chelipeds, they do not feed by grasping their prey but instead have long, hair-like fringes on their mouth parts through which they filter the water and extract food particles. These animals are to be found clinging tenaciously to the underside of stones and are very common. They may well be regarded as some of the most highly adapted animals on the rocky shore, as their flattened, broad shape enables them to fit into irregularities of rock surfaces almost as if they had been moulded.

Whereas the free-ranging animals are able to graze, forage or hunt, the fixed fauna has to make its food come to it. Often the fixed way of life involves animals in tremendous anatomical, physiological and behavioural changes—even extending to the larval or juvenile stages of their life cycles. Fixed animals obtain their food by one of two basic methods: extending their tentacles to grasp prey or filter-feeding. The fixed coelenterates extend their tentacles with their batteries of nematocysts and paralyze and grasp small organisms which come within reach. Sea-slugs, however, which graze on the tissues of sea-anemones, have developed a way of preventing the nematocysts from discharging their poison. Thus they can take in the nematocysts and transfer them to special areas on their backs, where they are used as second-hand (but still very effective) organs of defence by

I. LEWIS/NSP

these otherwise harmless and unprotected animals.

Most fixed animals, however, are filter-feeders, using the system for other physiological functions besides food-gathering, i.e. respiration, waste removal, expulsion of gametes or larvae. Sponges and several types of sedentary worms have different, sometimes very specialized, ways of filtering, which always involves generating a current of water, trapping solid particles and pumping out the waste water. Feather stars, some of the most beautiful animals on the shore, are also filter-feeders. When uncovered by the tide, they resemble a tangle of twisted red string but, when placed in clear water, their ten arms spread out, illustrating clearly the origin of the common name.

Acorn barnacles, trapped inside their hard covering, also employ a form of filter-feeding. The animals lie on their backs and, by thrusting out their feathery limbs through the hinged openings in the central doors covering their chamber, they are able to kick particles of food suspended in the water into their mouths.

Perhaps the most efficient and complex of all filtering mechanisms is that found in the tunicates, which pass enormous quantities of water through their sifting, plankton-net-like pharynx.

BANNISTER/NHPA

Feather stars like this tropical *Tropiametra* are the littoral representatives of the deep-water sea-lilies. Unlike their sublittoral relations, they are attached to the rock only as juveniles. The movements of the cilia on their beautiful, feathery arms, wafts detrital food to the central mouth.

Left: The lower areas of temperate rocky shores sometimes harbour numbers of squat lobsters such as *Galathea strigosa.* More closely related to crabs than to lobsters, they use their powerful pincers to grab their prey.

Some species of barnacles, such as *Balanus crenatus,* have to transfer sperm from one individual to another by means of a long penis. Because they are a fixed species, complex physiological adaptations ensure that they sett out of their planktonic larval phase within penis length of one another.

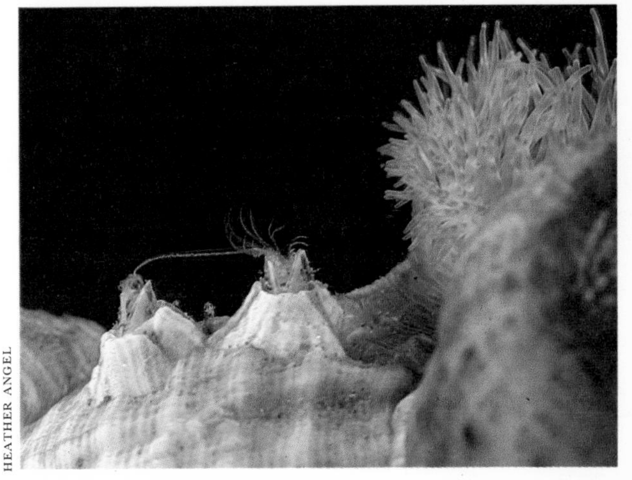

HEATHER ANGEL

## Reproduction and distribution

Reproduction on the seashore is not without its difficulties, particularly for the fixed species. The prevailing conditions are demanding enough for the adults of the species and would be even more so for the delicate larvae.

In many species, therefore, e.g. sea-squirts, echinoderms, worms, coelenterates, the larval stages are completed in the open sea, where conditions are more constant, and, on metamorphosis, the adult resumes its existence on the shore. This also helps to distribute the animals, preventing overcrowding.

In the fixed species, especially, fertilization is a problem. Most are hermaphrodite and usually occur in considerable numbers. Some, e.g. coelenterates and sea-squirts, can also reproduce asexually by budding. Generally, eggs and sperm are released in large quantities into the sea where external fertilization takes place. Some of the barnacles, however, transfer their sperm in spermatophores by means of an intromittent organ, which means that, on metamorphosis, they must sett within penis length of each other. The adults already attached therefore secrete a chemical substance to attract the metamorphosing larvae which thus setts close by.

In some filter-feeding fixed species, a form of internal fertilization takes place. Sperm is drawn in through the inhalant siphon and the eggs are fertilized within the animal or its tube. In the sea-squirt, *Dendrodoa grossularia*, the eggs are incubated in the exhalant chamber.

Another approach to the problems of reproduction on the shore is to lay fewer but larger eggs. As these eggs have more yolk the larva is comparatively well-developed when it hatches and better able to tolerate the conditions. Winkles and dog whelks lay their eggs in stalked vases or pouches, firmly attached to rocks on the lower shore.

Protection of the eggs may be achieved by brooding, a feature which is common in most shore fishes. The male pipe-fish has a ventral pouch in which the eggs are brooded until development is complete. The lump-sucker, *Lumpus lumpus*, which comes inshore only to breed, attaches itself to a rock by its ventral sucker and curls itself around its clutch of eggs, which are also firmly attached to the rock. Some amphipods also brood their eggs and the berried crabs and lobsters, with clusters of eggs attached to their abdomens are a familiar sight. In the splash zone, *Littorina neritoides* broods its eggs within its shell, liberating the young when fully grown, thus protecting them from desiccation.

## Microhabitats

Modern ecologists regard most major ecosystems as complexes of smaller habitats, each with its own micro-ecology. The seashore is no exception to this and such subdivisions are found on any shore.

Compared with particulate shores, the microhabitats of rocky shores are more conspicuous and, therefore, probably better known. Crevices in the rocks support complex and complete ecosystems, often with their own patterns of zonation, as do rock pools. The different types of rocks, as well as their position on the shore, influence the fauna which they support. The holdfasts of algae, particularly the larger species, are tortuously twisted and so provide many enclosed and protected spaces where small species can lurk. On very exposed shores, it is the holdfasts of the small coralline algae which provide the greatest number of niches. The numbers and variety of species supported by these holdfasts may be very great. In fact, almost any porous material found on the shore may constitute a microhabitat—compound sponges, for example, often contain extensive populations of worms.

Because of the more uniform nature of particulate shores, the physical conditions are less variable and there are fewer discrete ecological niches. However, the associations between animals on such shores are often far closer and more complex.

The seashore is a demanding environment, presenting an immense variety and complexity of habitats. It is because of this that the animals and plants show such a wide range of adaptation and form such a fascinating subject for study.

HEATHER ANGEL

## Shore fish

Whilst the majority of marine fish live in open ocean waters, some are adapted to the more restricted habitat of the seashore. Like many other littoral organisms, some of them have adopted a sedentary or nearly-sedentary habit and have evolved behavioural relationships with other shore animals and plants.

Like all organisms which inhabit a tidal area, shore fish have to be able to cope with periods of altering tidal cover and exposure. Some, e.g. the sand-eels, *Ammodytes*, bury themselves in soft sand, the trapped moisture of which enables the eels to survive and to respire. Sand-eels are morphologically adapted to burrowing—they are elongated, with narrow, pointed fore-quarters and a generally streamlined appearance which not only enables them to burrow easily but to be fairly agile even within the sand. The surface layers of a sandy shore may be used in a somewhat similar manner by stranded flat-fish. In the tropics, it is not unusual to find large numbers of tongue soles buried just under the surface of the sand during the period of low water.

Lack of tidal cover does not affect all shore fish

HEATHER ANGEL

Left: Many shore animals brood their young for a time. Crabs, like this female *Macropipus puber*, carry the egg mass under the abdomen, supported by specially modified limbs.

The pelvic fins of gobies are modified to form a sucker by means of which they can cling to rock surfaces. Here, a male *Gobbius paganellus* is guarding its mate's egg mass on the underside of a littoral boulder.

Rocky shore fish must be able to survive exposure to the air at low water. Many, like the tompot blenny *Blennius gattorugine*, are able to make limited use of atmospheric oxygen if they are stranded at low tide.

HEATHER ANGEL

adversely. Many fish are able to survive in the air for short periods—notably species which may have to supplement their oxygen supply in polluted or stagnant waters by gulping in atmospheric air at the surface. Some, e.g. the anabantids, have developed a special 'labyrinth organ' above the gills to enable them to do this more efficiently. *Anabas*, the climbing perch of tropical regions, is capable of surviving for prolonged periods out of the water and, in temperate waters, many of the gobies can survive exposure.

Air-gulping has led, in some shallow-water fish, to the production of sounds. By using a special modification of the swim-bladder, the toad-fish, *Porichthys*, is able to produce croaking noises. Primarily of use in mating, this sound can also be used to frighten off would-be attackers. Vocal power has been developed to a considerable level in the 'singing midshipman', *Porichthys porosissimus*, which, in the mating season, fills the air of parts of the American Atlantic coastline with its nuptial chorus.

A similar, but less well-developed air-breathing ability is found in many of the rocky-shore fish of temperate regions which, like invertebrates of rocky shores, attach themselves either permanently or temporarily to their substrata. An ability to use atmospheric oxygen is therefore a necessity during periods of low water.

Organs of attachment vary in form but they are almost invariably composed of modified pelvic fins. These fins may retain their fin-like form as in the gobies, or they may have lost almost all their outward fin-like appearance as in the suckers. The cup-like arrangement of the gobies' stiff-rayed pelvic fins can be extraordinarily effective in enabling these little fish to cling to rocks and thus withstand the scouring action of the tide. Used in conjunction with their pectoral fins the sucker also enables them to 'walk' over rock surfaces.

Sucker-fish of the family Discoboli show the most complete modification of ventral suckers. In the sea-hen, *Lumpus lumpus*, the oval pelvic fin sucker, which is partly encircled by the pectorals, enables the male to cling tenaciously to the rock where its mate's egg mass lies. *Lumpus* is well known as a faithful guardian—refusing to leave its charges despite attack by birds and rodents when stranded and held firmly by the sucker against violent water movements when inundated.

Almost equal tenacity is shown by the suckers of the littoral Cornish sucker, *Lepadogaster gouani*, and Montagu's sea-snail, *Liparis montagui*. The Cornish sucker has a flattened ventral region as well as a large ventral sucker and this, coupled with a generally low body-form, enables it to cling very firmly to solid surfaces. Montagu's sea-snail also has an extensive sucker but it is the relative flabbiness of its body which allows it to mould itself to the substratum.

Littoral algal cover is of great importance to shore fish. Apart from the protection which it provides against dessication at low water, it also provides a refuge against predators. In some species, such as pipe-fish and seahorses, the body-form has elongated so that when they are swimming vertically amongst the upright filaments of water-supported weeds, they are effectively camouflaged. A seahorse's prehensile tail enables it to cling to the protecting weed. A parallel adaptation is found in the tropical shrimp-fish, *Aeoliscus*, which swims head-down amongst the spines of long-spined sea-urchins and is thus protected and camouflaged.

The sticklebacks, well-known for their nest-building, use littoral weeds as foundations and raw materials. *Gasterosteus spinachia*, the fifteen-spined stickleback, makes its nest from pieces of filamentous weed bound together by a sticky kidney secretion. The weed is wound around the algae to make a globular structure several centimetres in diameter. By means of a complex behavioural ritual, the male fish entices a female through the nest where she lays her eggs. When he has fertilized them, the male guards the eggs until they hatch. Some shallow-water fish attach their eggs more directly to the larger algae. Gravid female dogfish swim amongst water-supported weeds and the pressure of the weeds against the flanks of the fish act as stimulus to laying. At each corner, the horny egg-case is drawn out into a long, coiled tendril which twists about the weed as the case is layed, anchoring the egg securely and holding it above the bottom in a current of water which provides a constant fresh oxygen supply for the developing embryo.

The food of shore and shallow-water fish may be very varied. Bottom-dwelling flat-fish such as flounders, plaice and soles, feed upon the invertebrates which abound in their muddy substratum—e.g. molluscs and worms. The rather more active gurnards eat the same type of food, supporting themselves above the bottom on their modified pectoral fin 'fingers'.

Most of the rocky shore fish are either scavengers, grazers or predators of small invertebrates. Some have developed highly adapted jaws and mouthparts to enable them to exploit particular food-sources. Thus, the trigger-fish and puffer-fish of tropical waters have strong 'rabbit teeth' at the front of their mouths to enable them to crush the shells of molluscs and the tests of sea-urchins. This type of modification reaches its extreme development in the parrot-fish which break off and crush pieces of coral head with their massive 'beaks'. Smaller and less dramatic perhaps, but no less elegant, are the delicate forceps-like mouthparts of the butterfly-fish, used for picking small organisms from amongst the folds and convolutions of the coral (Chapter 4).

Whilst not strictly a 'fish' at all in the taxonomic sense, the hagfish often occurs in shallow coastal waters. This elongated, snake-like animal, which is related to lampreys, enters the body of dead or

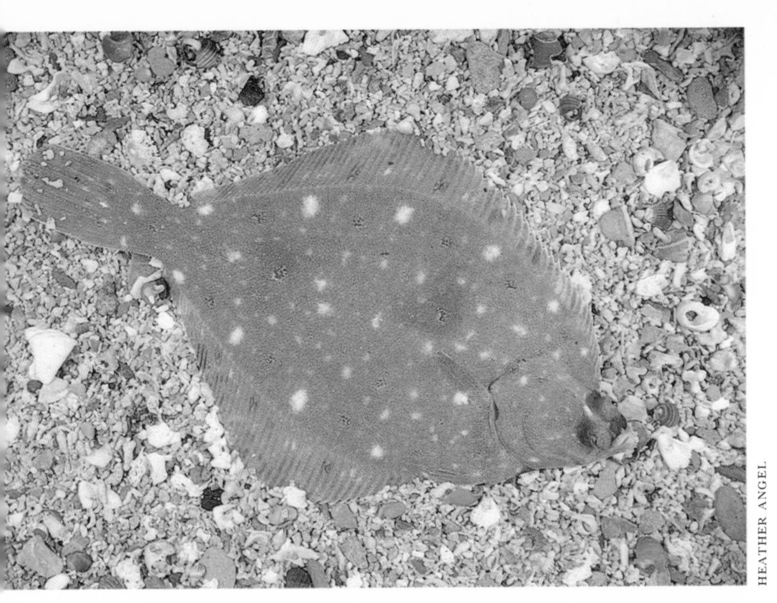

HEATHER ANGEL

The plaice (*Pleuronectes platessa*) is laterally flattened and lies with its left side on the bottom. To compensate for this, the larval left eye migrates over the skull to lie alongside its partner. These fish feed upon bottom-living invertebrates, such as worms and molluscs.

The subtropical and tropical seahorses (*Hippocampus*) have a prehensile tail which enables them to cling to upright weeds. Male seahorses incubate their mate's eggs in an abdominal pouch.

HEATHER ANGEL

dying fish, lubricating its passage with copious mucus which has given it its specific name of *Myxine glutinosa*. Once inside, it rasps away the organs, leaving its prey an empty bag of skin.

The associations which they form with other shore animals are, perhaps, the most interesting ecological aspect of the biology of shore fish. Many species are associated in a fairly simple manner involving complex modifications to only one of the partners, while others have much more complex relationships. Relationships between shallow-water fish and sea-anemones are particularly well known and the association of clown fish *Amphiprion* with tropical anemones is discussed in Chapter 4. Whilst it is complex, this association is certainly not so bizarre as that which exists between some of the cucumber-fish and sea-cucumbers. *Carapus acus*, an elongated, thin fish, enters the body of a sea-cucumber tail-first through its cloaca and takes advantage of the respiratory currents of its host to bring it planktonic food and oxygen. (It has also been suggested that the fish may feed upon the internal tissues of its host.) The gut is situated far forward in *Carapus* with the anus just behind the head. It can therefore defaecate by thrusting just the head and fore-quarters out of the sea-cucumber's cloaca without exposing the rest of its body. The fish may, occasionally, leave its host and re-enter—as indeed may other *Carapus*. If too many enter the same host, however, an excess of 'lodgers' may cause its death.

Whilst the majority of commercial fisheries exploit the fish populations of open ocean-waters, some inshore fisheries are sufficiently profitable to be commercially viable. Throughout the world 'sporting' fishermen seek their prey with rod and line or spear but in some areas—notably in the tropics—where populations are dense—marketable quantities of fish may be collected from the shore by these methods or by the use of simple cast-nets. In many tropical areas, due to the co-operative effort of several fishermen, the use of shore and boat-based seine nets produce greater catches of inshore species.

In temperate parts of the world small-scale fisheries are usually confined to plaice and dab fisheries in and near estuaries whilst the more commercially organized trawl for herring, sardine, pilchard, anchovy and mackerel in the relatively shallow continental shelf waters. The rich plankton of these shallow waters often makes fishing for the plankton-feeding species (such as the herring) a very profitable concern.

# Cora

Corals secrete massive, calcareous skeletons into which they contract almost completely when they are disturbed. When expanded, their basically anemone-like form can be clearly seen, as in this *Goniopora*.

# Reefs and Mangrove Shores

A casual observer travelling about the world as a tourist cannot fail to notice that in different regions, usually climatically distinct, different species of animals and plants occur. More scientifically, he will discover that the same species, when they occur in different regions, occupy different ecological niches and, conversely, that when basically similar niches occur in different regions, they are often populated by unrelated but similarly adapted species.

The science of zoogeography is concerned with this often climatically-linked variation in flora and fauna in different parts of the world. So far as marine biology is concerned, the major zoogeographically recognizable regions are: the two extremely cold zones—Arctic and Antarctic; the northern and southern cold zones—boreal and anti-boreal; north and south temperate zones—Atlantic and Pacific; Mediterranean; northern and southern sub-tropical; and tropical. These broad categories may be further sub-divided for particular biological disciplines and ecological purposes.

Many chemical and physical factors interact to produce the regional characteristics which have led to the evolution of geographically separate populations of marine organisms. Some, such as the great current and tidal systems of the oceans, are on a grand scale and may influence not only local marine populations but also the climate of the nearest land-mass. Others, such as the chemical content of the water may be of local and restricted influence—important perhaps to academic zoogeographers as a basis for subdividing the larger regions. The most conspicuous influence, however, and one which is often observable without special means, is temperature.

Bodies of water of different temperatures, moved by the great currents, can have a most profound effect upon animal and plant life. The animal life of the North and South Atlantic Oceans differs markedly in both composition and ecological organization. The North Atlantic is influenced by a complex of currents of widely different temperatures, such as the cold East and West Greenland Currents, the Labrador Current and the warm Gulf Stream. The Southern Ocean has the unique ring-shaped or annular currents of the West Wind Drift which circulate around Antarctica. As a result, the North Atlantic contains a patchwork of minor zoogeographical areas whilst the Southern Ocean tends to subdivide into annular zoogeographical bands which are also subject to complex ecological and physical layering at different depths.

Nowhere, however, is the effect of temperature more conspicuous than in the tropics and for this reason, as well as because of its rich ecosystems, it is a good example of a zoogeographical region. The tropical zoogeographical zone may be regarded as naturally defined in the same way that horizontal levels on a temperate rocky shore are defined by the zonation patterns of the algae. Here, the natural boundaries depend upon the largely temperature-controlled ecosystems which characterize it—mangrove swamp and coral reef.

Reef-building corals are sensitive to the temperature around them to a critical extent: the surface water temperature must not fall below 21°C if they are to survive. Permanent local lowering of surface temperature can prevent reefs from developing altogether. The westward-facing coasts of land masses have no coral reefs because the temperature of their coastal waters is lowered by upwelling cold water from the ocean depths. Mangroves, too, are sensitive to temperature variation, being unable to survive if the temperature falls more than 10°C from their ideal of 19°C.

Reef-building corals are restricted in their distribution to the area within 30°N and 30°S although they are restricted in the South Atlantic and occur in the North Atlantic as far north as Bermuda (32°N). Mangroves occur in the same areas but, unlike corals, they are able to grow on the westward facing shores of landmasses and may also extend outside the coral zone.

One of the first things a biologist visiting a tropical shore for the first time will notice is the greater number of species which occur there, as compared with temperate shores, where large numbers of individuals of relatively few species tend to be found. This state of affairs has probably arisen because of the high rate of primary production in the tropics. Good, strong, sunlight and high temperatures all the year round, with no seasonal variation, would lead to a plentiful and sustained food supply. This, coupled with a fast rate of biological turnover, would over evolutionary time allow many varieties to develop, varieties which would, eventually, become the many different species found on tropical shores today.

## Mangrove swamps

When mud particles carried down by river water meet sea water in estuaries, the suspended solids are attracted to one another and the clumped-together particles tend to sink, gradually building

Mangroves must ensure an adequate supply of oxygen to their mud-bound and water-logged root systems. *Avicennia* achieves this by means of aerial roots called pneumatophores, the walls of which are perforated by large pores.

The lower leaves of mangroves often act as a support for sedentary species. Here barnacles (*Elminius modestus*) cluster on the under-surface and edges of leaves of *Avicennia*.

Typical of the mangrove mud-fauna are the fiddler crabs (*Uca*). Males have one greatly enlarged claw which, although almost useless as a feeding tool, serves to attract a mate and, possibly, to delineate breeding territory.

up the mud banks that are so typical of estuarine shores. In the tropics the mud is almost invariably colonized by mangroves.

'Mangrove' is more a descriptive term than a precise definition, since species from several flowering plant families flourish in estuarine mud habitats. All mangrove trees and shrubs are characterized by the following adaptations: an ability to withstand the rotting effect of permanent immersion in waterlogged mud and the adoption of drought resistant characteristics in order to conserve water. This is necessary because although the plants are surrounded by water for much of the time, most of the water contains salts at a greater concentration than that of the sap of the roots and it therefore cannot pass into the tree. Rotting is usually prevented by the exaggerated development of spongy cellular layers around the roots. Some genera, e.g. *Rhizophora*, raise the main stem of the plant clear of the mud on stilt roots.

As well as its potential danger to the plant as a rotting agent, the mud could also cause suffocation, for green plants must be able to respire through their root systems as well as through their leaves. In order to make root respiration possible, some mangroves, e.g. *Brugiera* and *Avicennia*, have developed pneumatophores, vertical projections from the roots with numerous large pores (stomata) through which the roots maintain a snorkel-like contact with the air when they are uncovered by the tide.

These pneumatophores grow from usually extensive underground root systems. As a result, the establishment of even a few mangroves in a developing mud bank helps to hold it together. It may also serve to collect ever increasing quantities of flocculating mud and so add to the bank's size. This, in turn, allows more mangroves to become established and, in a surprisingly short time, an extensive forest will be formed.

Animal life in mangrove swamps is very varied, both in its composition and in its ecological distribution. Older areas may support large populations of amphibians, reptiles, birds and mammals but these have no direct bearing on the strictly mud/estuarine mangrove animals except, occasionally, as predators. In the more solid mud regions, which are essentially supralittoral, the fauna consists almost entirely of crustaceans, especially fiddler crabs (*Uca*) and grapsid crabs (e.g. *Seasarma*, *Grapsus* and *Plagusia*). Male fiddler crabs are easily identified because one claw, used for signalling and sexual display, is enlarged and brightly coloured. The colour of the claw and the way it is moved vary from species to species. The other claw, which is small and spoon-shaped, is used for feeding from the mud surface. The grapsid crabs, with their squarish, box-like carapace, may be seen scampering about amongst mangrove roots and in and out of burrows which are often excavated from the sides of enormous, pyramidal mounds of mud. These are not in fact built by the crabs, which are merely lodgers in the outer parts of the home of another, less conspicuous crustacean, the mud lobster (*Thalassina anomala*) which lives deep underneath the solid surface of the supralittoral mangrove, in the still waterlogged underlayers. This superficially lobster-like animal maintains its deep burrow by bulldozing through the mud, using its greatly enlarged chelipeds. The cleared mud is ultimately shovelled out of the burrow at the surface, forming the mounds which provide burrowing places for the grapsid crabs.

Although the supralittoral and splash zones of a shore are usually at that point which is horizontally furthest from the extreme low water spring tide level, this is only part of the case in a mangrove swamp. Landward from about mean tide level, a mangrove shore may be considered as developing in two directions—horizontally and vertically. As well as the normal horizontal supralittoral, another vertical one exists in the upper branches and leaves of the mangrove trees themselves.

At or about the extreme high water spring tide level, amongst the foliage of a *Brugiera* or *Sonneratia* mangrove, there are quite extensive populations of *Littorina melanostoma*, a gastropod beautifully adapted for a splash-zone existence. It migrates upwards as the tide rises, always keeping itself above high water level and grazing on algae and lichens on the leaves and bark of the tree. It can survive lack of contact with the water for long periods by closing itself into its shell with its operculum, sealing the boundary with secreted hardening mucus.

At about the low water neap tide level on the tree, that is on the lower branches, the undersurfaces of the leaves often bear a bivalve, *Enigmonia rosea*. From here down to the mud surface, exposed twigs, branches, stems, leaves and pneumatophores may also support heavy setts of barnacles. At the lowest level, the pneumatophores and stilt roots may carry permanently fixed bivalves such as *Isognomon* and *Chama*.

Most mangrove animals are mud-dwellers. The surface layer, like all muddy substrata, has a covering of diatoms and, usually, a rich deposit of detritus. Microscopic food material, especially in the shelter of mangrove roots, is eaten by collembolan insects (spring-tails). These animals, only a few millimetres long, may be present in such numbers that the surface of the mud looks quite grey/blue. In the same sheltered conditions there may be large populations of sea-slugs, *Onchidium*. More usually regarded as typical grazers of the diatom cover of littoral rocks, they may cover suitable mangrove mud almost entirely.

Like *Onchidium*, the other common surface molluscs are grazers. Perhaps the most typical of these are the ceriths, which range from small species like *Cerithidea obtusa*, which attaches itself upside-down to mangrove roots by byssus threads, to the massive *Telescopium telescopium*, which moves slowly about the mud.

In common with the animals of other fine-particle shores, most mangrove species live below the surface. The rich, humus-like soil which results from the decomposition of organic material from the mangroves, provides an ideal habitat through which worms and worm-like invertebrates can burrow, taking in mud and food as they go. Numerous bivalved molluscs inhabit the soft mangrove mud, though their distribution tends to be patchy and dictated by local conditions. The more massively shelled species may occur in quite stiff, often sandy soil, while those with more fragile shells and permanently extended siphons live a more sedentary life usually amongst soft, decaying organic material near the bases of the trees.

Some mud-living bivalves have special adaptations to the unstable conditions of the mud. The cockle *Anadara granosa* has a shell which is ornamented with ribs and projections which prevent it from sinking uncontrollably into soft mud and *Placuna placenta*, the almost transparent window-pane shell lies on its slightly-curved left valve, its large surface area acting rather like a snow-shoe.

Rooted eel grasses (*Enhalus*), in the lower mangrove shore, serve to stabilize the mud. As on the lower zones of sandy shores, these consolidate the mud and sand so that tube-living worms are often found there. The plant leaves also act as an anchoring point for algae and some animals, such as the ark shell *Arca auriculata*, which attaches itself by byssus. This often dense plant cover frequently provides shelter for sublittoral species such as the king crab *Carcinoscorpius*, especially for juveniles. Adult king crabs may travel up the shore as far as mangrove roots near mid-tide level. King crabs,

WALKER/NSP

King crabs, here *Tachypleus gigas*, are more closely related to spiders than to true crabs. They frequently come up onto the sheltered shores of mangrove swamps, feeding on molluscs and worms from the mud surface. Their respiratory system is unique amongst marine arthropods and their formidable looking caudal spine acts as a lever if the crab is overturned.

Mudskippers (*Periophthalmus*) are fish which spend more of their time out of the water than in it. Their common name refers to their habit of propelling themselves across the surface of the mud in repeated skips by lashing their tails. Since their pectoral fins are firm and somewhat splayed, they are also able to climb low branches of trees with comparative ease.

HEATHER ANGEL

related to the sea scorpions that lived in the oceans some two hundred million years ago, feed mainly on molluscs and worms, shovelling them up from the mud surface.

Since mangrove swamps are usually formed at the mouths of rivers, some animals which normally live in brackish water environments are often found there. This rather specialized part of the fauna may include amphibians and reptiles but is best known for the brackish water crustaceans and fish of the creeks.

Some of the crustaceans, such as the burrowing prawn, *Upogebia*, which burrows into muddy banks, may be said to belong to this habitat. Others should, perhaps, be regarded more as temporary, although long-staying visitors. Estuaries are almost invariably sheltered and so make ideal feeding and often breeding grounds for penaeid prawns and swimming crabs (e.g. *Scylla*), which often enter estuarine creeks in huge numbers on a rising tide.

Two genera of fish characterize the lower zones of the mangrove swamp; the mudskipper, which spends more of its time out of the water than in it and the archer fish, which spends most of its time in the water of the creek but catches its prey out of it.

The mud-skipper, *Periophthalmus*, is a form of semi-terrestrial goby which frequently lives buried in soft mud with only its mouth and goggling eyes protruding. To help it survive out of water, the mud-skipper's gill chambers can be filled with a mixture of air and water. Its skin has a horny layer but this, like the gill chambers, must be kept moist. Mud-skippers are best known for their habits of perching on the roots of mangroves and of skipping across the surface of mud or water—often for quite long distances. When perching, mud-skippers use their muscular pectoral fins for support and their pelvic fins are specially adapted to help them cling to vertical roots. This very common fish may occur quite a long way inland, wherever a muddy ditch or drain runs along the edges of old mangrove forest. In the breeding season the males establish territories and excavate nests in the mud, attracting a female to spawn by a leaping courtship display.

The archer fish, *Toxotes jaculator*, swims just below the surface of the water and ejects a jet of water or a stream of droplets from its upwardly pointing mouth at an insect sitting on a leaf or twig often up to 1·5 m above the surface. It is said to have excellent aim and to be capable of bringing down its prey successfully more often than not.

# Coral reefs

There is probably no ecosystem that so strongly evokes the atmosphere of the tropics as that of the coral reef. It immediately conjures up images of palm fringed beaches of silver sand, warm sea and jewel-like fish darting to and fro amongst quaintly-shaped coral heads. All of this is true, although there is much more to the biology of the reef ecosystem than this idyllic picture suggests.

Zoologically speaking, coral is a coelenterate, a close relation of sea-anemones, but one which secretes for itself a hard, limy cup (corallite) to surround and support its soft tissues. Its long evolutionary history has enabled it to diversify its form and structure to a considerable extent and it forms the basis of some of the most complex relationships between plant/animal and animal/animal that are to be found anywhere.

A first visit to a coral reef at low tide is likely to surprise those people who know coral only from museum displays. Although there may be areas of unbroken white where dead corallites and deposited calcium carbonate occur, the general effect is of a meadow where the muted browns, greens and yellows of the extended tentacles of corals, soft corals and anemones are punctuated sharply by the startling orange, red, yellow and blue of tiny fish.

The soft, plant-like colours of the corals are the first indications of one of the complex biological associations which characterize the reef. Within their tissues, the corals harbour millions of unicellular algae, known collectively as zooxanthellae. These contribute some of the products of their own primary food production towards the nutritional requirements of their animal hosts. Corals also catch food in a typically 'sea anemone' fashion, and it is now known that they can absorb some of the free, broken-down complex organic molecules directly from sea-water. This habit of using several different food sources is by no means confined to corals but is almost exactly mirrored in the giant clam *Tridacna*, which occurs on coral reefs and, to a greater or lesser extent, by many other marine organisms both within and without the tropics.

Because they are partially dependent upon symbiotic plants, reef-building corals are generally restricted to clean water where sunlight can penetrate and photosynthesis take place.

The chief structural difference between a solitary and a compound or colonial coral is that in the colonial form the original polyp subdivides repeatedly, giving rise to a number of organically-

NATHAN/NSP

At an estimated rate of 1 cm per annum, compound stony corals, cemented together by calcareous weeds, grow into extensive and massive reefs. The turbulent outer edge of this fringing reef is outlined in white breaker foam. To landward of the reef crest is the reef-flat, here covered with a shallow layer of water.

Many invertebrates live in close association with coral heads. This worm, *Spirobranchus giganteus*, has made its tube within a coral head and has thrust out its spirally arranged tentacles to trap floating food particles.

NEVILLE COLEMAN

connected polyps, each with its own cup-like corallite. The whole mass thus formed may be relatively small with individually large polyps (e.g. *Trachyphyllia*), relatively large with individually small polyps (e.g. the stag's horn coral, *Acropora*) or encrusting (e.g. *Platygyra* and *Porites*).

Many tropical solitary corals are large and massive. The coral family Fungidae, juveniles of which grow rather like inverted mushrooms on a stalk of corallite, reach a large size as adults. As they grow, the mushroom-like disc breaks away from its stalk and lies free on the sand on the reef flat where it may reach a diameter of over 25 cm or a length of over 1·5 m. It is the accumulation of corallites, alive and dead, of these often massive animals which forms the structure of a coral reef.

Whilst many colonial species are encrusting, adhering naturally to dead corallites below, much of the stability of a reef comes from plant material. The coralline algae *Lithothamnion* and *Lithophyllum* occur everywhere throughout the reef system, often growing as rock hard, dough-like lumps which, in suitable conditions, may spread into any available nook and cranny, cementing together much of the reef.

There are three forms of coral reef: fringing, barrier and atoll. All three, though different in overall form, are similar in construction. The basis of all reefs is an ancient bed of long-dead corallite material which may be of phenomenal thickness. Borings at Funafuti Atoll in the South Pacific at the end of the nineteenth century indicated a thickness of coral limestone greater than 340 m. Since then, seismic sounding at Bikini Atoll has indicated a depth of 2,000 m of coral material over the bed rock. Corals, however, only flourish in relatively shallow water, and the living reef occupies only the upper 55 m.

The warm, shallow water of a reef flat lagoon, to the landward side of a fringing reef, is ideal for the development of further coral colonies. In such a situation there is usually a complex assemblage of permanently fixed sponges, hard and soft corals, sea-anemones, tube-living worms and molluscs, together with a staggeringly wide variety of free-living species. It is here, too, that seaweeds such as the green coralline alga *Halimedia* find suitable anchorage on solid objects, their skeletons ultimately forming part of the coral limestone.

Soft corals and tropical sea-anemones often occupy a position in the ecology of the reef which is almost as important as that of the hard corals. When they are alive they cover large areas of the

BARNETSON/NHPA

The often jagged, interwoven outgrowths of stag's-horn coral (*Acropora*) frequently give protection to shoals of young coral fish, providing a safe habitat which larger predators cannot enter.

In addition to the hard corals—solitary and compound—which compose the reef, soft corals often abound. The long tentacles outstretched into the water actively filter out planktonic organisms.

BONE

reef. Some soft corals may form colonies which are over 1 m in diameter, each containing enormous numbers of polyps while tropical anemones may be large and very numerous. They do not, however, form any part of the reef's lasting structure since, when dead, they disappear entirely.

Giant clams feature prominently in the mechanical structure of the reef, since their massive, immobile shells can act as focal points where other calcareous material accumulates, particularly coralline algae. Giant clams vary in size from the small *Tridacna crocea*, about 12 cm long, which bores itself into coral rock, to species over 1 m long which lie on the reef surface. Like corals, giant clams have zooxanthellae in their tissues. These plants are known to be of the greatest importance to the feeding of the clams, which probably depend on them for nourishment at least as much as they do on the planktonic material which they filter from the sea-water.

The remainder of the fixed fauna of the reef, except for the sponges, consists of small species. Many of the sponges, too, are small, about 25 cm square but in the deeper water, just off the rim of the reef there may occur a giant among sponges—*Petrosia testudinaria*—bucket-sized and bucket-shaped. The central cavity of the sponge colony almost invariably supports its own miniature ecosystem of corals, small giant clams etc.

Some of the small fixed fauna, such as the soft-tubed worms and the sea-squirts contribute to the living economy of the reef but not to its permanent structure. Others, like the chalky-tubed worms and the bivalves which attach themselves permanently to the coral rocks ultimately join the other calcareous remains as part of the coral limestone. In the sublittoral part of the reef are two other groups of coelenterates, the false corals in which the hard skeleton is composed of spicules compressed together (e.g. *Tubipora musica*, the organ-pipe coral) and the sea-fans (e.g. *Gorgonia*) which have a central skeleton covered by the soft parts. These often beautiful animals form an important part of the fixed fauna.

Organisms which are insignificant visually may also be of importance to the general maintenance of a coral reef. Large numbers of Foraminifera occupy crevices and spaces in amongst the coral heads. Their dead tests gradually accumulate and, as they are overgrown by *Lithothamnion*, they, too, become incorporated into the cement which holds the reef together.

The conspicuous free-living animals of coral reefs are worms, crustaceans, molluscs, echinoderms and

FAIN/NSP

Bright colours and conspicuous patterns which may be disruptively camouflaging often occur in coral shrimps. Some species have developed complex 'cleaning' relationships with other reef animals such as anemones, corals and fish.

Coral reefs are complex ecosystems, involving many different types of animals. Some of the more conspicuous associated species are the giant clams *Tridacna*. The mantle contains vast numbers of microscopic algae which live symbiotically with the mollusc. The combined pigment of algae and mollusc colour the mantle tissues in bright shades of green, blue, red or yellow.

HEATHER ANGEL

Some harmless species have developed an association with more virulent members of the reef community. This porcelain crab (*Petrolisthes maculatus*) is protected from predators by the tentacles of the anemone *Radianthus*.

fish. The majority of the worms live hidden amongst seaweeds or buried in pockets of accumulated sand. When covered by the tide, they glide about amongst the fixed fauna, searching for food—usually dead and decaying plant material. In more sheltered areas, under coral boulders or amongst branches of living corals are many beautiful scale worms, e.g. *Chloeia*. Looking rather unlike other worms, they have paired flaps or scales which cover their backs and paddles tufted with glistening white bristles.

To a certain extent the crustaceans of the reef are rather like those of the tropical sandy or rocky shore. Pistol and mantis prawns occur in the reef-flat sand and crabs hide under coral boulders and amongst the weed. Several curious associations have developed, however, between some crustaceans and the dominant reef coelenterates.

Female prawns of the genus *Periclyemes* live amongst the stinging tentacles of the anemones *Stoichactis*. The tentacles provide superb protection for the prawn and the brood she may be carrying, while the prawn serves a useful cleaning function for the anemone.

Crabs of the genus *Cryptochirus* have an unusual relationship with colonial corals. The females take

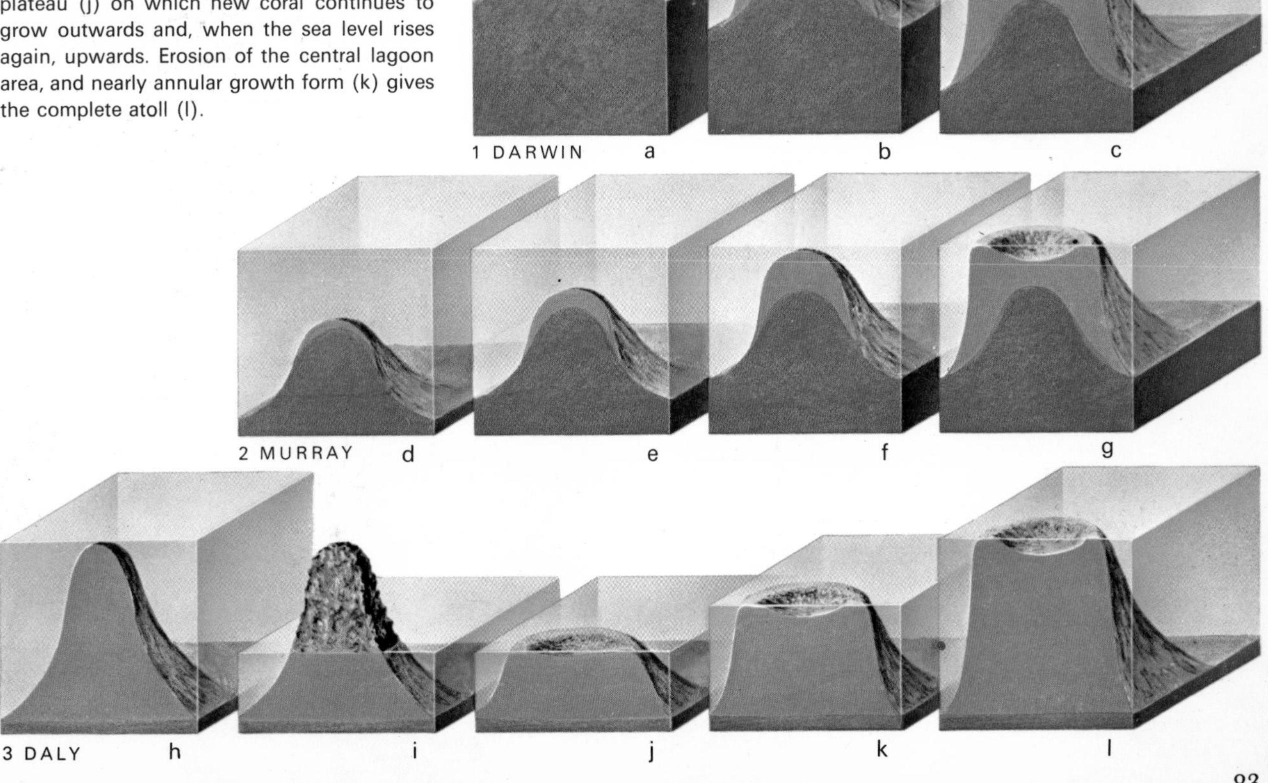


Three theories of atoll formation.
1 Darwin. A fringing reef forms around a solitary island (a). The island sinks geologically and the reef continues to grow, becoming an annular barrier reef (b). If the land mass continues to sink, the coral eventually becomes an isolated atoll (c).
2 Murray. A coral reef forms on a submarine mound (d). Geological uprisings (e) bring the growing reef near the surface (f). The living reef continues to grow out sideways and the central area is eroded, eventually forming an atoll (g).
3 Daly, and most modern theories. A large isolated coral reef (h) is uncovered and partly killed when the sea level falls during an ice age (i). The dead part is eroded leaving a plateau (j) on which new coral continues to grow outwards and, when the sea level rises again, upwards. Erosion of the central lagoon area, and nearly annular growth form (k) gives the complete atoll (l).

up residence in a cavity in a coral head when they settle out of their planktonic juvenile stage. They remain in the cavity which, as the coral continues to grow, closes around the crab until there is only a small opening connecting the inner chamber with the water outside. Water currents generated by the crab's movements bring in oxygen and food and, incidentally, keep the hole open by inhibiting coral growth. At mating times sperm transfer also has to occur through the opening in the coral since the male lives freely outside.

Barnacles of the genus *Pyrgoma* have an even more intimate association with coral. They cement their shell-like plates to a coral head when they sett out of the plankton. They then grow at the same rate as the coral so that ultimately the outer plates of the barnacle may be the only visible part of its calcareous structure, which is firmly cemented into the coral head and may be several centimetres long, extending through the coral colony.

Other crabs and prawns form associations with common reef species, especially with molluscs and sea-cucumbers. These will be described in the context of their host species.

The numerous species of coral reef gastropods include many which are beautiful in form, colour and in their adaptations to the requirements of their particular habitat. Many of these species are small and not easily seen, especially those which, like the family Magilidae, are associated with hard corals, soft corals or sea-fans. Of the common, conspicuous gastropods, the least spectacularly coloured are the turban shells (Turbinidae). These herbivorous snails, which are important to the economy of the reef as grazers of soft or filamentous algae, are chiefly notable for their characteristic and often colourful opercula. Gastropod opercula are usually more or less flat, horny or calcareous structures. In the Turbinidae, however, the operculum is massive, calcareous, often pigmented and, because it is protected by the snail's mantle, usually polished. The operculum of *Turbo petholatus* is a deep blue-green colour and is used, under the name of 'cat's eye', in jewellery.

If the opercula of the turbinids are jewel-like, the cowries can claim this distinction for their whole shell. Continuously protected by an overlapping mantle, the thick shell of these common and often abundant animals is polished to a porcelain-like gloss. Species of *Cypraea* feed on hydroid coelenterates and other encrusting plant-like animals while *Trivia*, which has a grooved, less glossy shell, browses among the coral on small, fixed sea-squirts.

NEVILLE COLEMAN

NEVILLE COLEMAN

BONE

Cowries are some of the most colourful of coral-reef molluscs. A living cowrie, such as this *Cypraea tigris*, presents a very different picture from the familiar glossy, speckled shells found in collectors' cabinets. Its fringed, warty mantle envelops it entirely, protecting its surface from abrasion.

Left: Brilliantly coloured starfish occur in all parts of the reef community clinging to coral and rock surfaces. The cushion star *Culcita schmideliana* lives in deep water where it grazes upon soft algae and detritus.

Left, below: Heavy-bodied wrasses with their thick lips and sharp incisor teeth are often encountered in the shallow water near the reefs. Some, like *Cheillinus trilobatus*, are highly coloured and patterned—probably disruptive against the dappled sunlight, reflections and organic colours of the reef. (About 45 cm)

Many of the small reef fish, like these squirrel fish (*Myripristis*), congregate into shoals, probably for mutual protection and for mating. (About 18 cm)

Amongst the other gastropod families such as the top shells, conches, frog shells, tritons, hairy tritons, murexes, vases, mitres, all of which contribute to the reef ecosystem, one stands out—if only for the way it preys on the invertebrates and fish which are its food. These are the cone shells (Cunidae). These often attractively-shelled snails capture their prey by firing a harpoon-like dart through their proboscis. The dart is coated with venom which is often virulent to man: *Conus textile* is particularly dangerous.

Apart from the giant clams, which are cemented into the coralline part of the reef, perhaps the most conspicuous bivalved molluscs are the fan shells (Pinnidae) of the sandy reef flat. They live buried in the sand, secured by a tuft of byssus threads. Their mantle cavity is often inhabited by pairs of pea crabs, *Pinnotheres*, which share the fan's food and in return presumably clean away detritus.

Four major classes of echinoderms are commonly found on the reef flat: feather-stars, starfish, sea-urchins and sea-cucumbers. The feather-stars (e.g. *Comatula*), which are filter-feeders, cling to the upper surfaces of coral heads and boulders to ensure easy spreading of their branching arms. Less delicate in appearance and more conspicuous on the reef

COLTMAN/SEAPHOT

surface are the starfish. Some, particularly those which live amongst and under coral heads, are small and hidden from casual view but scattered over the sandy areas there may be considerable numbers of large, brightly-coloured species, e.g. *Protoreaster nodosus*. This stiff-armed herbivore may be up to 30 cm in diameter and bright brick-red in colour.

The most famous of all coral reef starfish lives in deep water and therefore on the permanently submerged parts of the reef. This is *Acanthaster planci*, the crown of thorns starfish which is capable of stripping a coral reef of its living coral tissues at the rate of 1 m per *Acanthaster* per month. The population explosion of this species which changed it from a rare animal to an ecological pest was first observed in the mid 1960s and, since then, large areas of once flourishing reef have been severely damaged by it.

Reef-dwelling sea-urchins tend to be inconspicuous since, although they may have touches of bright colour they are usually well concealed. Many species are secretive by nature, occurring only in crevices or under coral heads. Others may hide themselves by sticking pieces of detached weed to themselves. One species which, despite its sombre colour, broken only by the white stripes on its dark purplish spines, is easily seen at the reef edge, is *Diadema setosa*. It is armed with long, thin, fragile spines which easily penetrate the skin and, having snapped off, quickly fester. The shrimp fish, *Aeoliscus*, often occurs amongst the spines, presumably cleaning away debris in return for protection, and camouflaged among the dark spines by longitudinal black stripes.

Less harmful to handle but equally if not more repulsive to most people are the sea-cucumbers which abound on the reef flat. The most noticeable are the really cucumber-shaped species such as *Holothuria scabra*, whose muscular body walls are quite stiff as they contain skeletal, plate-like spicules. These slow-moving animals feed on suspended detritus which they catch on mucus-laden tentacles surrounding their mouths. They have a fascinating method of self-protection. At the posterior end of the body are a number of tubular organs (the organs of Cuvier) which produce a very sticky secretion. When irritated, the animal secretes long threads which harden in sea-water, effectively enmeshing and immobilizing any attacker. Like the fan shells, large sea-cucumbers harbour species of pea crabs in their body cavity.

Less noticeable than the stiff-bodied sea-cucumbers are the synaptids, elongated rather serpentine echinoderms which may be up to 1 m

GRUHL/PHOTO AQUATICS

Most notorious of all coral-reef starfish is *Acanthaster planci*, the 'crown of thorns'. This beautiful and spectacular species is capable of destroying large areas of coral—it eats the soft polyps—and has highly venomous spines.

Some small reef fish, such as *Paramia quinquelineata*, live, for protection, amongst the long spines of the sea-urchin *Diadema*. The longtitudinal stripes on the fish enable them to blend effectively into the background of the spines.

PETRON/SEAPHOT

BENNETT/NSP

Sea-cucumbers often abound on the reef-flat. Here, *Holothuria argus* has discharged the sticky threads of its cuverian organs. The threads are used to entangle and trap would-be predators while the cucumber makes its escape.

NEVILLE COLEMAN

Many of the sponges which occur in deeper water off the reef-edge are massive and curious in appearance. The large exhalant openings or oscula of this scroll sponge are clearly visible.

long and are understandably mistaken for sea-snakes as they move sinuously amongst the corals. Their bodies are thin-walled and translucent. An easy way to identify a synaptid is to touch it: it will immediately stick to the skin since the scattered spicules in its body are sharp, anchor-shaped and protrude slightly through the body wall.

## Coral fish

The majority of coral reef fish are extremely beautiful to look at. Brightly, even gaudily coloured, patterned with stripes and spots, they dart in and out among the coral heads where sufficient sunlight penetrates the shallow water to make their colouration effective.

Colour in the animal kingdom is usually present for the purpose of attracting, camouflaging or warning. This may be linked with shape or anatomical sculpture designed to break up the outline of the animal or to make it merge into its immediate background. Coral reef fish display a variety of examples of these uses of colour.

Merging colouration and disruptive pattern are well shown in the bat-fish, *Platax*, and the leaf-fish, *Monocirrhus*. These resemble the dead leaves of

Coral fish are often very colourful, none more so than the butterfly-fish *Pygoplices diacanthus*. Bright banding of this kind is often used to distract the attention of would-be predators away from the sensitive head areas to less vital parts, like the edges of fins near the tail, where the pattern may be intensified. (About 15 cm)

GRUHL/PHOTO AQUATICS

PARISH

Disruptive colouration is clearly shown by *Microcanthus strigiatus*. Shoaling into groups tends to intensify the disruptive effect, further confusing potential predators. (Up to 10 cm)

Bright and often conspicuous colouration is displayed by many open reef fish from the mottled disruptive pattern of *Cleidopus* (right) to the canary yellow of *Ianthias* (far right) and the electric blue of *Paraplesiops* (right, above). Since *Paraplesiops* is a voracious carnivore, its starting colouration may be to warn other animals of its dangerous nature. (*Cleidopus*: 20–30 cm; *Ianthias*: about 20 cm; *Paraplesiops*: about 30 cm)

NEVILLE COLEMAN

PARISH

STARCK/PHOTO AQUATICS

mangroves, both in shape and colour, and thus are well hidden amongst the dead leaves in the shallow waters near mangrove swamps. They have also developed behavioural adaptations, giving the impression of floating, waterlogged leaves as they stalk their prey.

It is not so easy, however, to see the purpose behind some of the bright colours and patterns of many of the truly reef-dwelling species. In some, like *Chelmon rostratus*, *Zanclus*, the Moorish idols, and *Chaetodon ornatissimus*, the clown butterfly-fish, the vertical or oblique stripes probably serve a disruptive function against a backcloth of upright floating weeds, coral tentacles and vertical shadows. The angel-fish, on the other hand, use patterns of banded 'light and shade', e.g. *Pomacanthus imperator*, or a dappled effect which mimics the broken colour pattern of sunlight on troubled water as in *Holacanthus ciliaris*.

The common reef-dwelling damsel-fish are also banded but often with such bright colours that the protective purpose may be difficult to interpret. Some *Amphiprion* have transverse brown and white bands—a disruption which enables them to merge beautifully into a broken shadowy background. The well-known *Amphiprion percula*, on the other hand,

PARISH

Vertical striping, as in *Enoplosus armatus*, whilst disruptive in its own right, also serves a cryptic function when the fish swims amongst upright weed, or coral 'fingers'. (About 10 cm)

Many of the larger butterfly fish e.g. *Pomacanthus arcuatus*, have the large fin pattern which is necessary for stability in deep-bodied, laterally compressed fish. Both *P. arcuatus* and *Pomacanthodes semicirculatus* (right, below) have powerful crushing jaws with which they bite off and crush corals. The smaller fish apparently attached to *P. semicirculatus* is a cleaner wrasse (*Labroides*) which picks off parasites and other encrusting organisms. The wrasse's bright stripes are advertising colours used to attract the attention of its customers. (*Pomacanthus arcuatus*: about 30 cm; *Pomacanthodes semicirculatus*: up to 30 cm; *Labroides*: about 8 cm)

Right: Trigger-fish, e.g. *Balistapus undulatus*, are so-called because of the lockable, erectile spine on the dorsal fin. This is apparently used for wedging the fish into coral crevices when it is asleep, or hiding from predators. (About 25 cm)

KOPP/PHOTO AQUATICS

DEAS/SEAPHOT

PHOTO AQUATICS

is banded with white and bright orange—whilst this can hardly enable this species to merge with a shadowy background, the startling contrast very effectively breaks up the outline of the fish.

Similar breaking-up patterns of bright colours are often found in the trigger-fish, e.g. *Balistapus* and *Rhinecanthus* and in the related file-fish, e.g. *Alutera* and *Melichthys*. These fish inhabit cavities in and near coral heads and clumps of algae and their patterns, although bright, effectively destroy their 'fishy' outline against their background.

Some of the most beautifully shaped and disruptively coloured fish in the reef community are the shrimp-fish of the genus *Aeoliscus*. They live amongst the spines of the long-spined sea-urchins, *Diadema*, where, due to their longitudinal black and white banding and their habit of swimming head-down, they merge completely with their spiny surroundings.

Not all reef fish use colour for camouflage or disruption. Some, e.g. *Pterois volitans*, the lion or dragon-fish, use their bright, conspicuous colouration as a warning since their spines are very venomous. They are usually found in deeper water off the reef edge, with the dangerous moray eel. The poisonous-fleshed puffer-fish, e.g. *Arothron citrinellus*, and the related tobies, e.g. *Canthigaster margaritatus*, have brightly coloured warning patterns which are used, in conjunction with self-inflation, to frighten off would-be assailants.

Much more pacific in their advertising colouration are the 'cleaner wrasses' *Labroides*. These little fish, which remove parasites etc. from the skin and gill-chambers, even from the mouths of larger fish, advertise their presence by means of conspicuous black, blue and silver longitudinal stripes. This, coupled with a complex of specialized movements, enables would-be 'customers' to recognize them and so refrain from eating them as they perform their sanitary functions.

The cleaner-wrasses also provide, indirectly, an example of another, more sinister use of colouration. In the same localities as the cleaner-wrasses live the related sabre-toothed wrasses, e.g. *Aspidonotus*. These voracious carnivores have almost identical colour patterns and movements to *Labroides* and are thus able to approach their prey with impunity and bite chunks out of them.

Perhaps some of the most curious examples of 'hoodwinking' colour patterns found in the animal kingdom are those which cause would-be attackers to concentrate their efforts upon less critical areas of their prey. A technique which has been evolved in some of the butterfly-fish is to camouflage their eyes by means of coloured or black banding or by black patches and to produce pigmented eye-spots on the trailing edges of dorsal or ventral fins near the tail. In this way, blennies of the genus *Runula* which attack the tissues near the eyes of the butterfly-fish are tricked into biting a less vulnerable area.

Several species depend upon the coral for food. Moorish idols graze on the algae which are everywhere throughout the reef. While they are peaceful and rather defenceless, the related surgeon fish, *Acanthus*, are certainly not so. Near the tail of these elegant and delicately coloured fish are a pair of scalpel-sharp spines which can be unsheathed from the protective pocket if danger threatens. Surgeon fish, too, graze on algae and have specially adapted

DEAS/SEAPHOT

HACKMAN/PHOTO AQUATICS

Right, above: Groupers such as this *Epinephelus* are often abundant in the deeper parts of the reef community. Because of their size and gastronomic qualities they are often hunted by man. (1 m or more)

Few fish so faithfully live up to their sinister appearance and reputation as the moray eel (*Muraena*). Typically lurking in a coral crevice, its head shows the protruding nostrils which are linked to its highly sensitive olfactory system. (Up to 2 m)

NEVILLE COLEMAN

The red snapper (*Lutjanus sebae*) is an abundant fish over Australian reefs. The head stripe obscures the eye but its reddish colour, shown up by the photographer's flashlight, will appear black in natural conditions when there is little or no red light. (Up to 1 m)

PETRON/SEAPHOT

Puffer and porcupine fish (*Diodon hystrix*) inflate themselves with water to frighten off attackers. Their larger size, which makes them both intimidating and difficult to swallow, is made even more alarming by the sharp spines which stand out from the skin. (Up to 30 cm)

GRUHL/PHOTO AQUATICS

Surgeon-fish, e.g. *Acanthurus monroveae*, have a pair of razor sharp spines which can be retracted into sockets just in front of the tail. The incisor teeth of these common reef fish are flattened and rasp-like at the edge, to enable them to graze encrusting algae off rock surfaces. (About 40 cm)

Tropical chromid species are usually found in schools, swimming above coral reefs, feeding on microscopic zooplankton and copepods. (About 12 cm)

GRUHL/PHOTO AQUATICS

FAIN/NSP

Reef fish often suffer from the unwelcome attentions of the razor-toothed blennies (*Aspidontus*) which, disguised as cleaner fish, attack the body instead of the parasites. To direct their attention away from the delicate tissues around the eyes and gills, many species, like *Chaetodon capistratus*, have a pattern of lines which lead to the less vital area around the trailing fin edge and tail. A pigmented eye-spot completes the deception. (Up to 18 cm)

Below: The parrot-fish, e.g *Scarus sordidus*, have a pair of massive teeth in each jaw and large, pharyngeal teeth in the throat. With these they bite off and crush whole pieces of corallite from which they digest the soft parts. These often colourful fish spin a nest of mucous threads for protection when they sleep at night. (About 35 cm)

GRUHL/PHOTO AQUATICS

teeth with flattened, crenellated edges for rasping the minute plants from the coral rock surface.

Parrot-fish, e.g. *Scarus*, also have specialized teeth. Here each jaw contains only a sharp-edged plate forming a bony 'beak' which gives them the ability to crush pieces of coral so that they can digest the soft tissues. They also have massive grinding teeth in their throats to help to pulverize the food.

Butterfly-fish have teeth that are either chisel-like (*Chaetodon*) so that corallite and polyps may be cut off together, or forceps-like (*Forcipiger*). The forceps-like teeth are probably used to nip off individual polyps or to probe among the coral crevices.

Many of the reef fish have developed complex behavioural associations with the reef coelenterates, including the coral itself. The abundant and ubiquitous damsel-fish often live among the larger reef anemones. The most famous, *Amphiprion percula*, the orange and white-banded jewel or anemone clown fish, lives amongst the tentacles of anemones of the genus *Stoichactis*. Like certain other species, it seems to be able to move among the sensitive stinging capsules of the anemone without triggering them off. *Amphiprion*'s scales are covered in mucus, and it may be that this contains a substance which desensitizes the trigger mechanism of the stinging cells so that they do not 'fire' even when they are stimulated.

One of the most curious associations is that between trigger-fish (e.g. *Balistapus* and *Rhinecanthus*) and coral. Trigger-fish have a long bony spine at the leading edge of the dorsal fin. This spine is erectile and can be locked in its upright position by a smaller, second spine. The purpose of this

PARISH

Bottom-dwelling species like the goat-fish (*Upeneichthys porosus*) search carefully over the sandy surface of the reef-flat, moving sensory barbels under the jaw to detect the presence of the small invertebrates on which they feed. (30–40 cm)

NEVILLE COLEMAN

Left: The stone-fish (*Synanceja verrucosa*) is one of the most dangerous animals on the reef-flat. A powerful venom is secreted and transmitted along the stiff spines on its dorsal fin and gill-covers. So closely does this species match its background of stones and coral debris that it is almost impossible to distinguish it as it lies in wait for its prey. (About 25 cm)

The clown fish *Amphiprion* are often very numerous in the shallow water-area of the reef-flat. They are able to prevent coelenterate stinging cells from discharging and are, therefore, often found amongst the tentacles of reef-flat anemones. (About 10 cm)

BONE

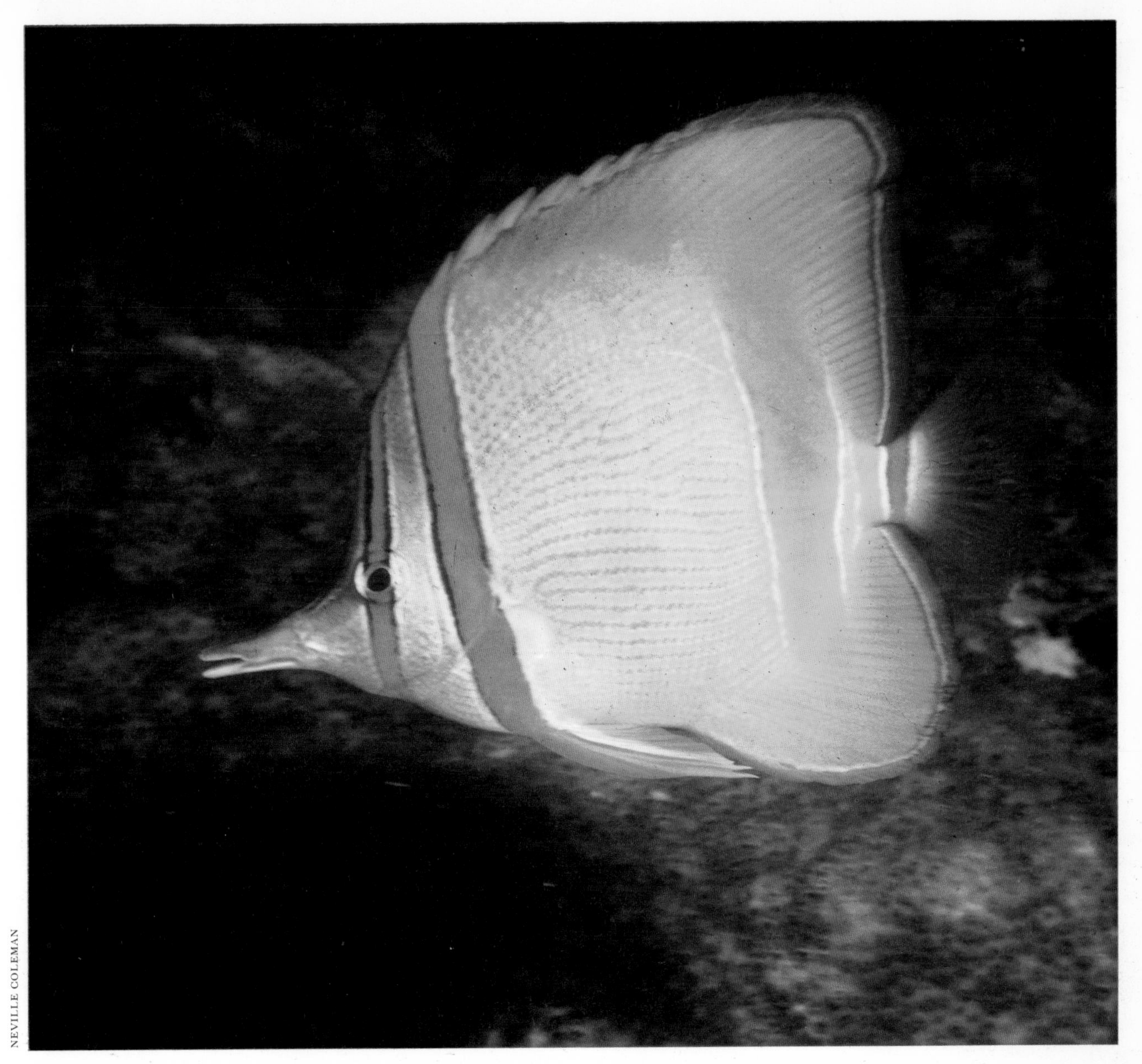
NEVILLE COLEMAN

The long snout of *Chelmon marginatus* enables it to probe amongst the convolutions of coral heads in search of small invertebrates to supplement a diet of nipped off coral polyps. (About 15 cm)

curious mechanism was for a long time a mystery but it is now known that the fish uses it to wedge itself in a cavity in the coral either to sleep or to avoid attack. The spine can be locked in position so rigidly and the fish can wedge itself so firmly that even very considerable force will not dislodge it.

Although fish such as surgeon and trigger-fish are well protected there are few predators to attack them in the waters of the reef. Blennies, moray eels and the occasional shark are probably their main enemies. One of the ugliest and most dangerous of fish, the stone-fish, *Synanceja verrucosa*, uses its venom for defence only. It lies motionless on the reef floor, camouflaged by its sculptured, stony appearance. If it is trodden on, the sharp spines on its heavy skull eject a poison that is usually fatal to human beings.

Whilst the tropical zone is no more important to the great complex ecosystem of the ocean than any other zoogeographical region, it is certainly unique in many respects. Its specialization, the number and variety of its species—both on shore and in the coral seas—its associations, its colour and the curious and often bizarre modifications of many of its inhabitants, combine to give it a biological ambience possessed by few other regions.

# The Plankton

Plankton is the general term applied to the small animals and the microscopic plants that float freely in midwater, carried passively by the swirls and eddies of the water currents. The animals are called zooplankton and the plants phytoplankton. The phytoplankton requires sunlight to grow and survive and, in consequence, it is almost entirely restricted to the surface 100 m. This is in contrast to the zooplankton which occurs at all depths. The plants are the original source of all food in the sea, as well as much of the oxygen in the atmosphere. Both the energy, which is contained in the food, and the oxygen are produced by photosynthesis.

## Phytoplankton

On land the plants are large and obvious; in other words there is a large standing crop. Similarly, in shallow fringing seas, large seaweeds and submarine flowering plants, such as turtle and eel grasses, grow in profusion. These plants have a large standing crop that can be seen and measured. However, seaweeds and marine grasses need a substrate on which to anchor; in the open ocean there is no substrate since the sea-bed is thousands of metres below. One exceptional seaweed does occur in the open ocean and that is *Sargassum*, which sometimes occurs in large quantities floating at the surface. The major producers of plant material far out to sea are the species of the phytoplankton—tiny single-celled plants which range in size from 1–500 μm (1 μm = 1 thousandth of a millimetre or 1 millionth of a metre). In sunlight these microscopic plants grow and divide extremely quickly, especially if they are well supplied with fertilizing mineral nutrients. So rapid is its growth that, if unchecked, the phytoplankton would turn the sea into a thick soup in a few weeks. However, the zooplankton responds quickly by growing and soon grazes down the plants. These herbivorous animals are the link in the food chain between the phytoplankton and the carnivores; the latter include commercially exploited species like the herring.

The grazing of the herbivorous animals keeps the standing crop of phytoplankton very low, although a litre of water may contain hundreds of thousands of tiny plants. Scientists measure how much carbon dioxide is taken up by phytoplankton in water samples, to assess the rate of plant growth. The amount of carbon dioxide 'fixed' is a measure of the primary production i.e. how much the plants grow per day. In the open ocean each year about 25–100 mg carbon are fixed under each square metre of the sea's surface. In the richer waters on the continental shelves the production is higher. For example, in the North Sea, the annual production is 100–250 mg carbon fixed per square metre, which is equivalent to the amount of plant growth occurring on pasture land. When these figures are multiplied up to estimate the total annual plant production for all the world's oceans, a figure of $2 \times 10^{16}$ g is obtained. This enormous quantity of plant production represents the use of only 0·2–1 % of the sun's energy arriving at the sea surface.

## Types of phytoplankton

The smallest of all the phytoplankton types are the μ-flagellates, some of which are only 1–2 μm across. Their submicroscopic size (even some bacteria are larger) means that they have to be studied using electron microscopes. They present great problems as to how they can be preserved, of how to identify them, and how to assess their ecological significance. They appear to be a heterogeneous group whose common feature is the possession of one or two long hair-like locomotor structures called flagella. Many may turn out to be the transient stages of larger, more familiar forms whose complex life cycles are unknown. It is thought that they are ecologically most important in open oceanic water; as much as 75 % of the chlorophyll in open oceanic seawater passes through filters with pore sizes of 10 μm.

The coccolithophores are a little larger and are better known, probably because of their palaeontological importance. Their fossil record of minute calcareous plates (coccoliths) stretches back 500 million years into the Cambrian era. Chalk is the massed deposits of coccoliths laid down in the Cretaceous era. Each coccolithophore cell has an outer covering of these coccoliths, which vary in size from 1–35 μm and have a delicate crystalline structure of calcite, a form of calcium carbonate. Coccolithophores are predominant in tropical seas, especially where the water is impoverished in nutrients. However, occasionally so-called 'chalky water' occurs in the Norwegian Sea; this is due to blooms of *Coccolithus huxleyi* in concentrations of up to 100 million cells per litre. These blooms are eagerly sought out by the herring fisherman, as they usually indicate water conditions favoured by the fish, which means that fishing is likely to be particularly good. In the Mediterranean, coccolithophores have been reported from very deep water, well below the photosynthetic compensation depth. These cells must either be dormant or else be able

The sailor-by-the-wind (*Velella*) floats at the surface of the sea with its sails projecting above the surface. Within its tissues, plant cells called zooxanthellae live symbiotically. The zooxanthellae are relatives of free-living plant plankton.

Pages 98–99
Plankton is the term used to describe the animals and plants that drift passively in the water currents. This large jellyfish *Rhizostoma,* swims gently along, driven by the pulsations of its bell. Accompanying it is a shoal of larval fish, seeking the protection of the jellyfish's presence.

to survive by utilizing the minute concentrations of dissolved organic compounds. There are cells, called olive-green cells, which are a regular feature of depths below the euphotic zone. They are too small for even the group to which they belong to be identified. Since they cannot be cultured or adequately preserved, electron microscope studies have not helped. These mysterious cells are abundant enough to be exploited by some of the deep living planktonic animals.

The diatoms are perhaps the most important group of phytoplankton. They flourish most abundantly in upwelling areas where the water is rich in dissolved silica. They utilize the silica to manufacture their glassy cell wall which is known as the frustule. The surfaces of the frustule are etched and ornamented with ribs and pores. Under the electron microscope, the frustule is revealed as consisting of geodesic dome and rib structures which give it maximum strength combined with maximum lightness. In highly productive areas, such as in the region of the Antarctic Convergence, accumulations of the frustules form soft muds of diatomaceous oozes on the sea-bed. Geological deposits of these oozes are mined commercially for a variety of uses, ranging from the manufacture of polishes and toothpaste to dynamite.

The other phytoplankton group of major importance is the dinoflagellates. They straddle the boundary between plants and animals. All have the cellulose cell walls that are typical of plants and many are photosynthetic. However, in some species, the ability to photosynthesize can be lost and the cells then live heterotrophically, using the dissolved organic compounds in the water as their energy supply. Still others are holozoic, feeding on particles and even small animals. *Noctiluca* is a large dinoflagellate, nearly spherical and a millimetre in diameter; it is a serious predator on fish eggs. *Noctiluca* is often the organism responsible for the flashes of bioluminescence that occur in the surf in late summer in temperate latitudes. Many other dinoflagellate species have this ability to produce flashes of light.

Quite a number of animals contain dinoflagellates called zooxanthellae, which live symbiotically within their tissues. These zooxanthellae occur in the reef-building corals, several of the surface-living animals, such as the sailor-by-the-wind (*Velella*), and in the mantle of the tropical giant clam, *Tridacna*. Blooms of some of the planktonic dinoflagellates can be extremely spectacular, often discolouring the sea. Some of these 'red tides' result in the release of powerful poisons produced by the dead and dying plant cells. In 1947, in a massive kill, 150 kg of dead fish were washed up along every metre of the west coast of Florida. In 1971, 75% of the shag population of the north-east coast of Britain was wiped out. Finally, in 1799, 150 people were killed in Alaska by paralytic shellfish poisoning; shellfish contaminated by a dinoflagellate bloom are extremely toxic.

The other phytoplankton groups are less important ecologically. The silicoflagellates were abundant in the geological past, but are so no longer, although they are widespread. Likewise the blue-green algae are believed to have been one of the earliest forms of life on Earth. Few are of much significance today but there is one extremely important species, *Trichodesmium*. This is a filamentous species, in which the filaments adhere together into loose bundles. Vast blooms of this species occur, especially in nutrient-poor tropical seas. These look like a reddy-brown scum which accumulates in slicks or windrows. They cover thousands of square kilometres, and gave rise to the name of the Red Sea because they occur frequently there.

## Zooplankton

The zooplankton has a far more extensive size range than the phytoplankton, from single-celled protozoan animals a few hundredths of a millimetre across to little crustacean relatives of crabs several centimetres long. The large animals, which are more powerful and able to swim against the currents, are called nekton. The division between nekton and plankton is ill-defined; for example, large jellyfish 2 m across the bell drift passively in currents and behave like plankton although they are much larger than many nektonic animals. The classification is usually one of practical convenience: plankton is caught in small fine meshed nets, whereas the nekton is caught in coarser meshed trawls.

## Larval forms

The zooplankton contains many animals that are only temporary members. These are larval forms, some of bottom-living animals, some of deep midwater animals, and others which grow up into large nektonic species. In shallow shelf seas the larvae of the bottom-living animals may outnumber all the other types of plankton. Worms and shellfish in their breeding season produce myriads of tiny eggs which hatch into transparent delicate little larvae. These larvae have some ability to swim, powered by many tiny whip-like structures called cilia (cilia are shorter, finer and much more numerous than flagella). The larvae feed on phytoplankton and grow into quite elaborate forms. The growing larva gradually becomes too heavy for its simple ciliary swimming to keep it up in the water and it settles on to the bottom. There it metamorphoses into the adult form and starts life in its new habitat. The longer the larva's life in the plankton, the further away from its parent it is likely to settle. The function of this larval stage is to disperse the species as widely as possible so that every favourable habitat can be colonized.

Many commercial fishes have planktonic eggs and larvae; the cod, for example, has planktonic larvae, although it is a bottom-feeding fish. The survival of the larvae is vital to the good recruitment of young fishes to a fishery. The problem of how to predict good recruitment is still unsolved. It is important to ensure that there is a large enough stock of breeding fish. However, good feeding conditions for the larvae seem to be more important for survival than a large number of eggs being released into the water. The tiny size of many larvae necessitates their feeding on phytoplankton; anything larger would be too big for them to handle or swallow. As they grow there is a time when they switch from being herbivores to being carnivores or omnivores. In bottom-living animals this often coincides with their settling on the bottom and metamorphosing into the adult form. This is a critical stage in the larva's existence and, when attempts are made to culture animals, this is the stage at which most larvae die. Midwater animals with herbivorous larval stages either have to lay buoyant eggs, which will float up to the surface, or make spawning migrations into the upper layers. Below the photic zone, where there is no phytoplankton, larvae either have to feed carnivorously straight away or feed on detritus dropping down from the surface. The animals may either lay fewer

This mixture of animal plankton includes larvae of crabs, which when they mature will live on the sea bed, and 2 mm long orange and yellow copepods which are permanent members of the plankton. These animals eat even smaller phytoplankton.

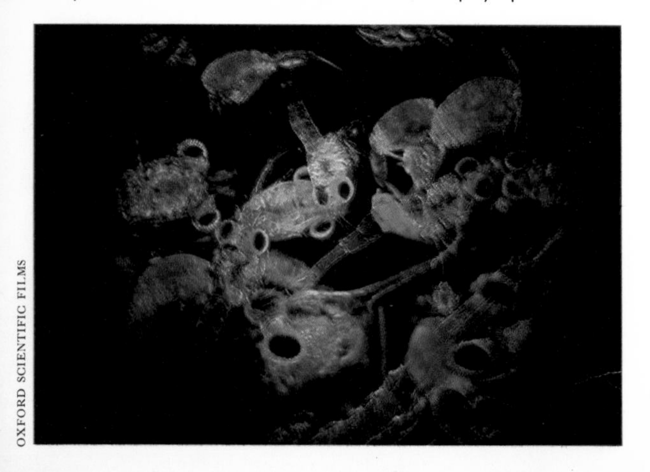

OXFORD SCIENTIFIC FILMS

The large lobes of the foot of the veliger larva of a snail are covered with tiny cilia. These cilia beat continuously to keep the steadily growing larva with its increasingly heavy shell afloat in the plankton.

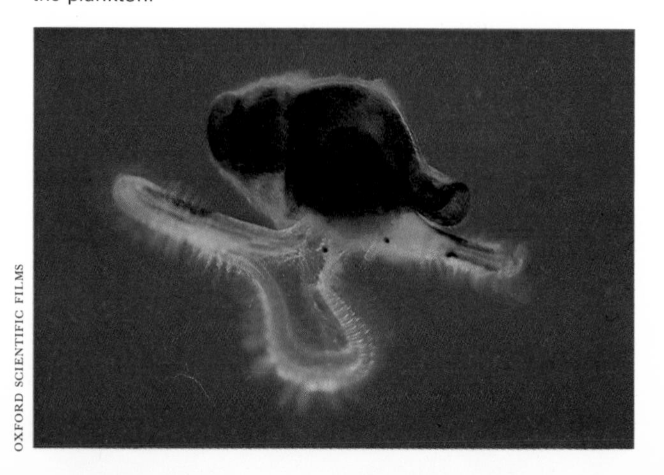

OXFORD SCIENTIFIC FILMS

but larger eggs, so that the hatching larvae are larger, or they may brood their offspring. As the small herbivorous larvae in the surface layers grow and change to a more carnivorous diet, so they sink deeper into the water. Thus many animals change their depth distribution throughout their life history; this is called ontogenic migration.

## The permanent plankton

Animal plankton occurs at all depths of the ocean, although it is most abundant near the surface and gets progressively rarer as the depth increases. The composition of the plankton changes with depth, but representatives of most planktonic groups occur at most depths. These groups range from single-celled protozoan animals, relatives of amoebae and ciliates like *Paramecium*, to jellyfish, worms of various types, a variety of crustaceans, molluscs and relatives of sea-squirts called salps.

**The Protozoa** Many of the protozoans are very difficult to preserve and identify, so they are very little studied. Their importance in the oceanic eco-system is probably far greater than is realized at present. For example, in the surface hundred metres a group of ciliates called tintinnids occur very abundantly. They feed on the tiniest of the phyto-plankton cells, which are too small for most other animals to catch. These protozoans are probably most important in linking the decomposer bacteria back into the food web and recycling organic material back through the system so that it is not lost by sinking down to the bottom. Any carcase of a small crustacean is rapidly invaded by a seeth-ing mass of tiny ciliates so that, within a couple of days, they completely clean out all vestiges of the carcase's soft parts.

Many of the planktonic larval forms are very bizarre. This larval cranchid squid has its eyes on short stalks, and its tentacles are still developing. In the centre of the tentacles is the mouth, with its parrot-like beak.

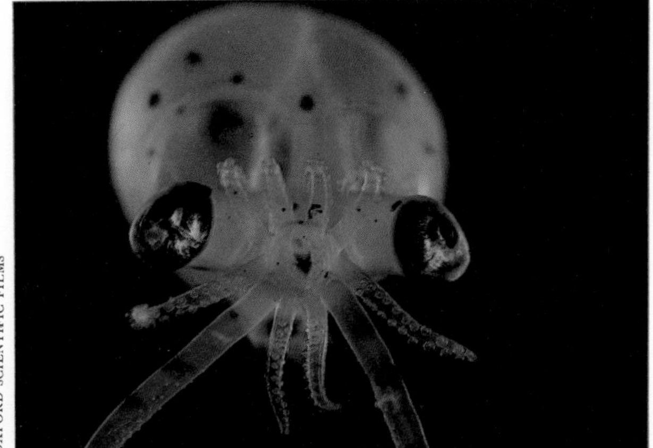

OXFORD SCIENTIFIC FILMS

Far better known are the rhizopod Protozoa, which are all amoeboid animals. This is because many of them produce skeletons, which are found in the soft muds or oozes on the sea-bed and so are of considerable interest to palaeontologists. The calcareous skeletons of the Foraminifera are im-portant indicators of what the sea-surface tem-peratures were in the past. In *Globigerina* the direc-tion in which the shell coils reverses should the water temperature increase above 5°C. Immense areas of ocean floor are covered by the soft greenish accumulation of millions of these tiny skeletons, which is known as Globigerina ooze.

The Radiolaria are protozoans with skeletons of most exquisitely sculptured silica. There are a great variety of shapes, but one of the most common is spherical with radiating long, fine spines. Some radiolarian and some foraminiferan species have been shown to contain plant cells inside them. These plant cells are symbiotic, providing the protozoan with nourishment in exchange for pro-tection. These species, to keep their symbionts healthy, must live in the sunlit waters of the euphotic zone.

**Jellyfish** Jellyfish belong to the group called the Coelenterata, which also includes sea-anemones and the siphonophores. The siphonophores are relatives of the most notorious coelenterate of all—the Portuguese man-of-war, *Physalia*. *Physalia* has a large float which looks rather like a large light bulb floating at the surface. Below the float hang long highly retractile tentacles that stream far beyond the rest of the animal. It is on these tentacles that occur batteries of the little stinging cells that are typical of all coelenterates. The cells discharge their poisonous darts if anything brushes against them. The poison can rapidly kill a fish and will even paralyze human beings if they are badly stung. Siphonophores used to be thought of as colonies of individuals called polyps, each with an anemone-like structure. Some of the polyps are modified into little muscular swimming bells, rhythmically pulsating and squirting water out to push the whole animal through the water. Other polyps are modified for feeding, others for repro-duction, others as protective bracts and yet others as long tentacles. All these polyps are arranged along a central stolon in a special order. As none of the polyps can survive independently the whole organism is no longer thought of as a colony but as a complex individual. The gas-filled floats of

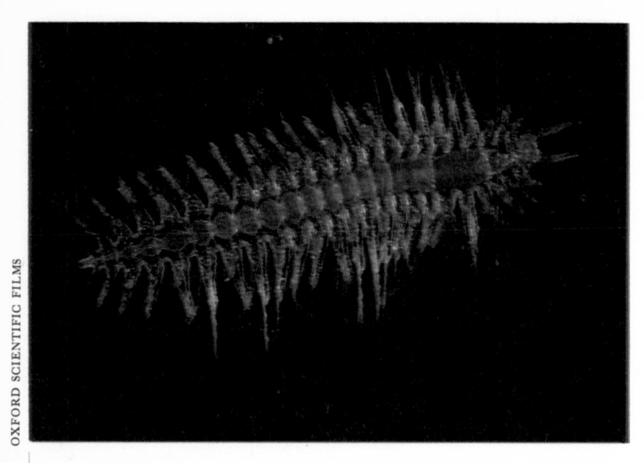

OXFORD SCIENTIFIC FILMS

Many of the planktonic polychaete worms appear deceptively delicate but they are all active carnivores. Like many other species living in the near surface layers this animal is extremely transparent, its only colour is the orange of the eyes.

OXFORD SCIENTIFIC FILMS

This euphausiid shrimp *Meganyctiphanes norvegica* is one of the important herbivorous planktonic animals in the North Atlantic, where it is an important item in the diet of whalebone whales. It is a close relative of Antarctic krill.

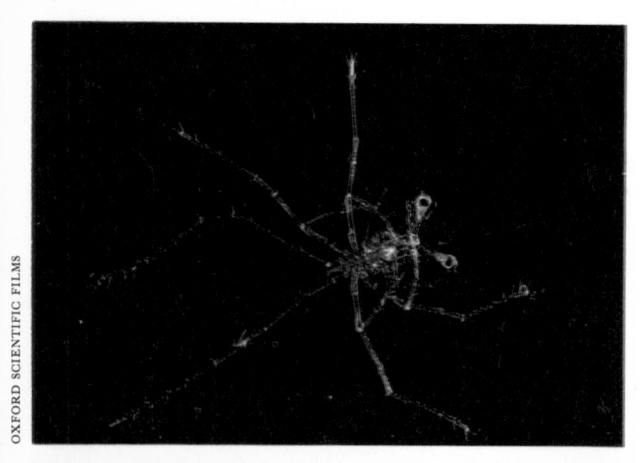

OXFORD SCIENTIFIC FILMS

Many species of lobster go through this larval stage, called a *Phyllosoma.* Most grow to about 10 cm across the legs but giants over 30 cm across occur. These feebly-swimming herbivores eventually grow into active carnivores.

some of these siphonophores reflect so much of the sound output of echo-sounders that they produce a false bottom echo or deep scattering layer.

The more familiar jellyfish are predominantly coastal because most species have a stage of their life cycle which is attached to the bottom. They can slowly swim through the water by pulsations of the bell. Fine tentacles fringe the edge of the bell and coarser ones surround the mouth in the middle of the underside. Most are carnivores and can kill quite sizable fish. The most poisonous is the sea wasp, *Chironex fleckeri*, which causes the deaths of several bathers each year off the northern coasts of Australia. A few species of jellyfish feed on fine particles which they trap on mucous sheets that are moved over the bell by fields of cilia. The moon jellyfish, *Aurelia aurita*, is an example of a fine-particle feeder and is a familiar jellyfish in coastal waters throughout temperate seas. It still possesses small stinging cells, but they are too weak to sting either human beings or other animals. Close examination of a live plankton sample will often reveal dozens of tiny jellyfish that are the dispersal stages of bottom-living hydroids.

**Worms** Some of the more abundant planktonic animals are the arrow worms (Chaetognatha). These torpedo-shaped worms have lateral fins and a flat tail fin. The head is armed with two rows of long curved mouth spines and, although few species grow more than 5 cm long, they are very important predators. They sense movements of their prey and dart forward to grab them with their mouth spines. In the North Sea, the chaetognaths probably eat more newly hatched herring larvae than any other type of animal. The herring larvae that survive get their own back later on, as they consume large numbers of the arrow worms.

A few species of true worms are permanent members of the plankton. They have segmented bodies, with each segment being expanded into paddle-shaped projections down each side of the body. They swim by serpentine movements but with the waves passing forwards from tail to head. Most species are extremely transparent and some have huge elaborate eyes on their heads. One of the commonest genera, *Tomopteris*, is interesting in sometimes having a tiny species of jellyfish living parasitically inside its blood system.

FAIN/NSP

*Porpita* is a relative of the jellyfish. Its central disc is chambered and gas-filled. Below the disc are delicate blue tentacles which surround the mouth. *Porpita* feeds on smaller animal plankton than the larger Portuguese man-of-war.

FAIN/NSP

Left: The Portuguese man-of-war (*Physalia*) is also a relative of jellyfish and sea anemones. Its gas filled float rides on the surface and acts as a sail. The long highly extendible tentacles carry stinging cells that paralyse fish that touch them.

Left, below: The group of jellyfish that is most dangerous to man is the Cubomedusae. The sea-wasp (*Chironex fleckeri*) causes several deaths a year in Australian waters, but all the species have powerful and painful stings. This species shows the characteristic four groups of tentacles.

Trawls from 1,000 m in the North Atlantic often contain many of these deep living jellyfish *Atolla*. They grow up to 30–40 cm across the disc. The mouth is on the underside and is fringed with reddish-purple tentacles and four pairs of whitish gonads.

NEVILLE COLEMAN

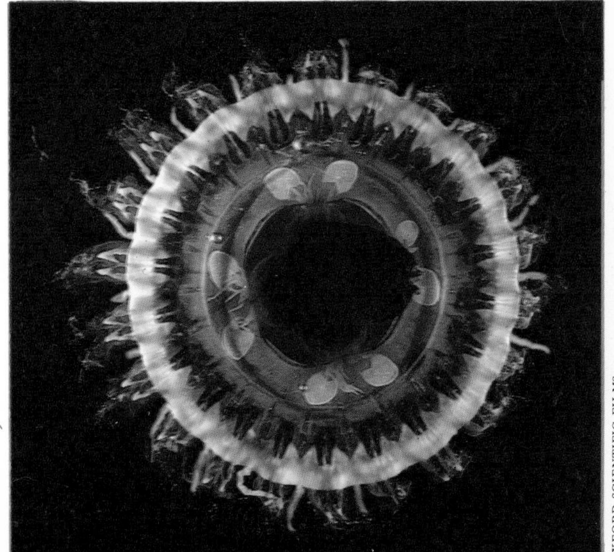

OXFORD SCIENTIFIC FILMS

**The Crustacea** Many of the large crabs and lobsters have planktonic larvae, but there are several less well known crustaceans that are extremely important in the plankton: the water fleas, the copepods, the ostracods and the euphausids. They all possess two pairs of antennae and three pairs of limbs modified as mouth parts, and thoracic and trunk limbs which vary in number between the various groups. The limbs are jointed and carry an assortment of spines and bristles. Each limb may be specially adapted for either swimming, manipulating food, grabbing prey, or sieving fine particles out of the water.

Most abundant of the planktonic crustaceans and, for that matter, of all plankton, are the copepods. They occur at all depths, from right at the surface down to the bottom of abyssal trenches. There are many hundreds of kinds but at any one depth only four or five species may be really abundant. Copepods range in adult size from less than a millimetre to over a centimetre. The smaller species which abound in the sunlit surface layers are predominantly herbivores. The larger species that usually occur deeper down are carnivores or detritus feeders. Many animals show a tendency to grow to larger sizes in cold environments and so it is not surprising to find the bigger copepods below 1,000 m and at high latitudes.

The first copepod ever described was *Calanus finmarchicus*, one of the most important planktonic species in the North Atlantic. It usually feeds by sieving phytoplankton out of the water. It generates vortices of water with its mouth parts. The vortices swirl through a mesh of fine bristles (setae) and the trapped plant cells are scraped off into the mouth.

The Norwegian fishermen used to call the masses of *Calanus* in their coastal waters krill. When whaling began in the Antarctic the name krill was given to another type of crustacean, the euphausid, *Euphausia superba*. These shrimp-like animals, up to 25 mm in length, swarm in immense shoals and are the staple diet of the baleen whales. The Russians are now trying to exploit these huge concentrations as a source of protein, both for human consumption and for feeding stock. *Euphausia superba* is a herbivore feeding on the exceedingly rich production of diatoms available during the Antarctic summer. Relatives of the krill occur at lower latitudes; although many are herbivores, others are omnivores or carnivores.

Ostracods are little bivalve crustaceans that are now known to be far more important in the plankton than previously believed. Microfossil remains of the shells of the bottom-living species have been studied by geologists in deposits originating in all eras back to the Cambrian. They are used, for example, in the study of oil-bearing deposits. In the Indo-Pacific ostracods are well known as the producers of a brilliant bioluminescence.

The planktonic ostracods have thin shells that are too fragile to become fossils. Most of the species have no eyes and so cannot see images. An exception is the largest ostracod of all, *Gigantocypris*. The largest specimens of this nearly spherical animal are up to 35 mm in diameter. It has enormous reflectors attached to its naupliar eyes, which take up nearly a third of the anterior part of the body and are specially adapted to see bioluminescence. *Gigantocypris* is also unusual for an oceanic ostracod in retaining its eggs inside the valves of its shell and brooding them until the larvae are developed enough to be able to fend for themselves.

Other planktonic crustaceans that brood their eggs and young are relatives of the water fleas, *Podon* and *Evadne*, both of which belong to a group called the Cladocera. These little animals, less than a millimetre long, suddenly appear in the plankton, build up into huge swarms and then apparently disappear. During the build-up of the swarm, reproduction is asexual and parthenogenetic—i.e. the eggs develop without being fertilized by a male. The eggs and larvae are brooded and released only when well developed.

A different type of brooding occurs in *Phronima*. This relative of the sand-hopper (amphipod) is a carnivore that feeds on gelatinous organisms like siphonophores and salps. The female cuts the centre out of one of these to make herself an open-ended barrel. She sits in the centre of the barrel and jet propels it through the water by the respiratory current produced by the beating of her abdominal legs. The eggs are laid onto the walls of the barrel and the young larvae hatch and grow, feeding on the food carried into the barrel by the inhalant current. Many of the other amphipod species are intimately associated with species of jellyfish. Juveniles of *Hyperia galba* live commensally on the underside of the moon jellyfish, *Aurelia*. The juveniles of other genera are external parasites ectoparasites on siphonophores. Some of the oceanic species which live in the euphotic zone have some extraordinary developments of the head and the eyes. In *Streetsia*, the cylindrical elongation of the eye takes up a third of the body

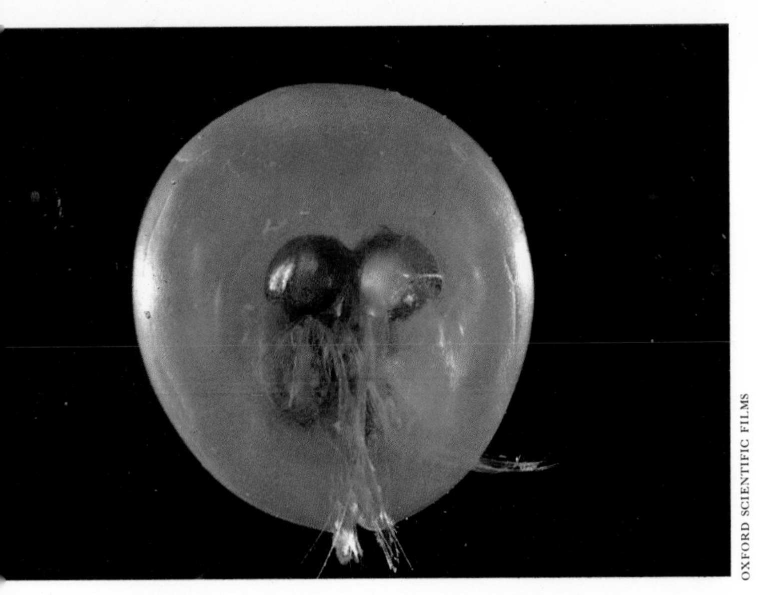

OXFORD SCIENTIFIC FILMS

The largest of the bivalve ostracods is *Gigantocypris*. Living at depths of around 1,000 m, the huge reflectors of the eyes are particularly well adapted to perceive the light produced by other animals at these otherwise darkened depths.

*Phronima* is an amphipod that carves out a barrel for itself from the bodies of gelatinous animals such as salps. Clinging to the inside of the barrel, *Phronima* jet-propels itself through the water. The barrel is also used to protect the developing young.

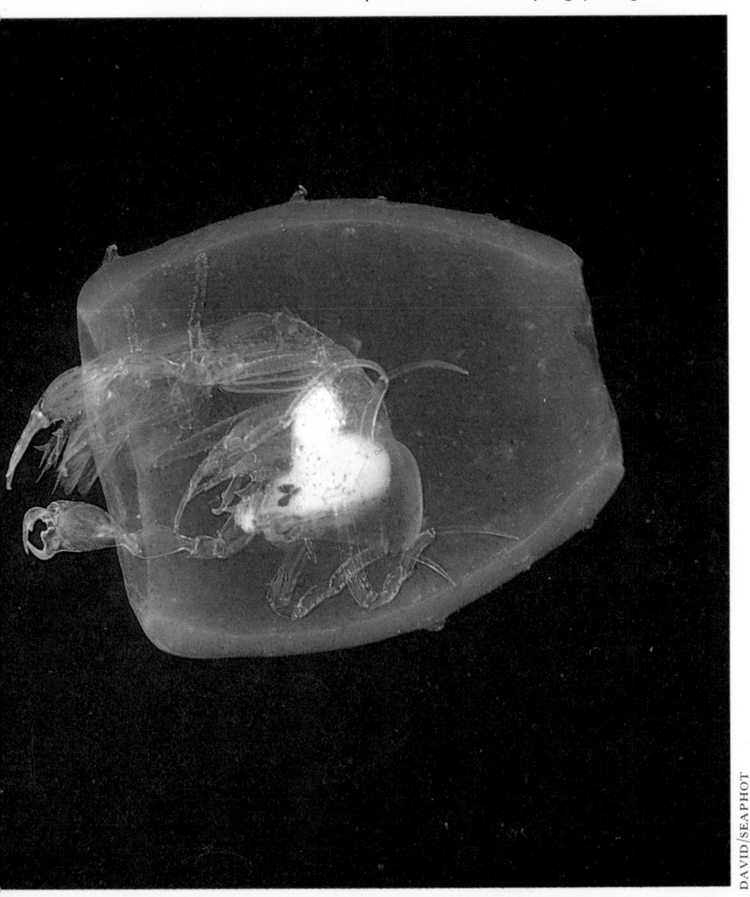

DAVID/SEAPHOT

length; there is no satisfactory explanation as to why the eyes are so fantastically developed in these animals.

**Salps** The salps which are preyed on by the amphipod *Phronima* are planktonic relatives of the tunicates (sea-squirts), sedentary bottom-living animals. Their bodies are cylindrical with openings at front and back. Rhythmic pulses of the muscles round the body wall squirt water through the body, so that it is sieved through a fine mesh-work bag formed by the gills. The gills take up most of the body space and have tiny hair-like cilia which slowly move a sheet of mucus over the gill surface. The mucus acts as a sieve, which retains even the smallest phytoplankton and bonds chemically with organic molecules dissolved in the water. Salps, when swimming, cannot help but feed; when they are satiated the mucus, instead of being consumed, is pelleted and passed straight out undigested. These faecal pellets may be an important route by which plant production from the surface layers is transferred rapidly to deep depths. Salps are interesting also in having an alternation of generations. The asexual individuals produce a long chain of budded-off sexual individuals whose fertilized eggs give rise to new asexual individuals. They grow extremely rapidly and, with this very rapid form of reproduction, huge populations can appear in a couple of weeks. In this way they can exploit to the full short-lived but luxuriant blooms of phytoplankton that occasionally occur.

*Pyrosoma* is a colonial salp in which the individuals are arranged peripherally round a hollow cylindrical matrix. The colonies can attain a length of over a metre. Living within the colony are bacteria that produce a steady and continuous bluish glow of luminescence. The turtle is one of the few animals that has been reported as feeding on *Pyrosoma*.

**Molluscs** There are a number of relatives of the snail that are permanent members of the plankton. The largest group are the pteropods, which are sometimes called sea-butterflies. The foot of the snail has been developed into two wing-like structures. Few pteropods are bigger than a centimetre long so the vigorously beating wings propel them quite quickly through the water. All over the surface of the wings are fields of cilia. These slowly move mucus sheets across the wings to the mouth, where the sheet with phytoplankton stuck to it is eaten. Some species have delicate beautifully sculptured shells. In areas where they abound, the bottom muds are solid with empty shells.

The pteropods without shells are carnivores. The mouth everts into a proboscis which is armed with teeth; the only prey of one shell-less pteropod is a shelled species. Another group of predatory planktonic molluscs are the heteropods. These too have evertable proboscises armed with sharp teeth.

There are one or two planktonic sea-slugs. *Phillirhoe* is a little brown-coloured sea-slug with its body studded with light organs. Attached to the head of some specimens is a small medusa. At first biologists thought the medusa was a parasite, but studies of the life history of the sea-slug proved the reverse. The larva of *Phillirhoe* attaches itself parasitically to the medusa. Eventually it outgrows its host but remains attached to it. Another curious sea-slug occurs floating at the surface: *Glaucus* stays afloat by periodically swallowing air. It feeds on coelenterates that live at the surface such as the sailor-by-the-wind, *Velella*. Like shore species, as it feeds it manages not to discharge the stinging cells in the tentacles of its prey. In its gut, special amoeboid cells take up the stinging cells and carry them via the blood to the outer surface of the body. There, along the bunches of finger-like projections (cerata) on the side of the body, the stinging cells are arranged so that they can then be used in the defence of the sea-slug.

Another voracious feeder on the surface-living siphonophores is the purple sea-snail *Janthina*. It has a bluish-coloured shell and keeps afloat by secreting a bubble float that looks just like a cigarette end floating on the sea. The snail also lays its eggs on the float and these hatch directly into tiny snails which crawl away on the surface film. *Janthina*, when it is attacked, releases copious clouds of a purple dye.

## Plankton production

It has been estimated that, each year, about $2 \times 10^{16}$ g of phytoplankton are produced in the world's oceans. Sedimentation rate on the sea-bed occurs exceedingly slowly and consists mostly of dust blown off the land and the calcareous and siliceous skeletons of some of the animals and plants. So virtually all the organic material that the phytoplankton produces is eaten by herbivores. About 10% of the food eaten by the herbivores is turned into new growth and this is termed their ecological efficiency. The other 90% or so is 'burnt' up by their respiration, in maintenance activities and as energy used in swimming and feeding by the herbivores. The figure of 10% is possibly an underestimate, since so many fish are caught in the North Sea that the ecological efficiency needs to be 15–16% if the measures of phytoplankton production are correct. However, assuming that the 10% figure is roughly correct, then the world production of herbivores is $2 \times 10^{15}$ g. This suggests that the maximum catch of fish that can be taken from the sea each year is in the order of 100 million tonnes. Clearly the herbivores must be extremely efficient in catching the plant cells, especially as in regions like the centre of the North Pacific, 75% of the plant material (biomass) will pass through a sieve with 10 μm mesh size. The herbivores are thus extremely efficient filter-feeders.

## Filter-feeding

The really small herbivores, such as many protozoans, are able to feed on individual cells of the smallest plants. Similarly copepods and euphausids will seize individual large diatom cells, but most phytoplankton that is eaten is sieved out or trapped. The finer sieves are usually formed from mucus sheets. The mucus can trap minute particles and it can chemically absorb large molecules in solution from the water. In the impoverished areas of ocean, where the smaller forms of phytoplankton predominate, those animals using mucus sieves are particularly important.

In richer areas, where the water contains high levels of nutrients, the larger plant cells start to make significant contributions to the available standing crop of the phytoplankton. Animals with coarse sieves, such as the copepods and euphausids, then become the most important herbivores. In the copepod, *Calanus finmarchicus*, adult females have filters with a mesh size of 5·7 μm, the larvae have finer filters with a mesh size of 3·2 μm in the younger and 3·8 μm in the older larvae. Any particle retained on the filters is consumed indiscriminately, so the larvae avoid competition with the adults for food by eating finer particles. Experiments with a surface-living species of copepod, *Acartia*, have shown that it can effectively clean the water of large-celled phytoplankton 16–35 μm across, but flagellates with a size of 6–8 μm were filtered out very much more slowly. An adult *Calanus* can clear 70–100 ml of water of phytoplankton in a day, but can also see an individual large cell over a millimetre away, move towards it and seize it.

## Patchiness

Filter-feeding involves the expenditure of quite considerable amounts of energy. So it is not surprising to find that the herbivores stop feeding if the phytoplankton is not abundant enough for them to actually gain by feeding. The minimum concentration of phytoplankton needed for a herbivore to start feeding is called the threshold concentration. The threshold concentration for most herbivores is above the average concentration of phytoplankton in the surface layers, so unless the zooplankton is to be in a permanent state of near starvation it must exploit patches of high concentrations of the phytoplankton.

At scales of over 1 km in upwelling regions, or during the spring bloom in temperate latitudes, the reproductive rate of the phytoplankton is fast enough to maintain a patch in the face of its continual dispersion by horizontal diffusion. In regions where the phytoplankton is dividing more slowly the patches need to be 10–100 km across.

## Vertical migration

These patches of concentrated phytoplankton probably attract zooplankton into them. Most plankton vertically migrates up each dusk and down again at dawn. The animals relate the timing and the extent of their migration to the rate of change of light intensity. In an area of low productivity the water will be much clearer and the migrations tend to be much more extensive. The differences in both speed and direction of water currents are greater the deeper the animals swim down, so the next night they will ascend back into a totally different body of surface water. In regions where the production is high, the water is more turbid and daylight penetrates to a much shallower depth. The vertical migrations are less extensive, so the animals tend to resurface in much the same body of water. Indeed rich feeding conditions may inhibit diurnal vertical migration in some species.

The extent of the migrations is surprising; euphausids, and even copepods and ostracods only a millimetre or two in length, make regular daily excursions of 250–500 m vertically. As the sun sets a great upward movement starts. Some species halt in midwater, some at the thermocline, but others move right to the surface. These migrations are not restricted to the plankton; many nektonic species perform similar migrations. One little lantern fish swims up into the surface from depths of over 1,250 m in the North Atlantic off the Canaries. During the night the animals may scatter down and this is called the 'midnight sink'. At first light there is an initial concentration of the population back towards the surface, followed by a rapid descent into the depths as the sun rises.

Numerous explanations as to why the animals should migrate have been put forward but none is considered to be fully satisfactory. One of the first suggestions was that the animals moved down to escape from predators hunting by sight. Another was that the animals moved down to avoid toxic blooms of phytoplankton; this suggestion was then inverted to suggest that they vertically migrate to seek out blooms of phytoplankton. An interesting suggestion is that, since deeper down the water temperature is cooler, the animals move to depths where their metabolism is slower and more efficient, so saving energy. Another effect of cooler temperatures on animals is to make them produce fewer but larger eggs. Large eggs have greater viability and so the chances of survival of the hatching larvae are increased. The theory of conserving energy is certainly not true in all species, since the slowing effect of the lowered temperature on the animal's metabolism is counteracted by the acceleration effect produced by increasing the pressure.

None of these explanations is entirely satisfactory since they do not explain why some species migrate at certain stages of their life history but not at others. Alternatively only half the population may move up on any one night. A large number of species are non-migratory, or their migratory behaviour may alter seasonally. The way migration may alter seasonally is shown by the common Atlantic copepod, *Calanus finmarchicus*. This species overwinters in deep water as a non-feeding juvenile, living off its food reserves of fat. In the spring they mature, start to feed and vertically migrate again and then breed. The Pacific counterpart, *Calanus plumchrus*, also overwinters as a juvenile, but matures and breeds without ever feeding again. At the beginning of winter, 70% of the animal's dry weight is lipid—either fat or waxes. Marine organisms use these lipids both for their short term and long term reserves. This contrasts with terrestrial animals, in which considerable amounts of their reserves are in the form of carbohydrates particularly one called glycogen. Chemical analyses of planktonic animals show that they contain only minute quantities of carbohydrates.

Lipids are so extensively used by marine organisms that it has been estimated that half the organic matter synthesized by phytoplankton is converted at some time into fats or waxes.

## The structure of the food web

As the world's human population continues to grow there is a continued increase in the need to exploit the oceans as a source of food. A full understanding of plankton ecology is a primary requirement to prevent yet more mistakes in managing these resources. The structure of the food web is extremely important in determining the crop that can be taken without damaging the resource.

We have already seen that upwelling areas are centres of big fisheries in the tropics and that the largest fishery of all is in the region of the Peru Current. There the anchovetta feeds directly on the rich blooms of phytoplankton by sieving the cells out onto its gill rakers. This little sardine-like fish misses out the plankton link in the food chain except during the El Niño. This is during February to April when a warm current pushes south stopping the upwelling. The anchovetta switches its diet to copepods. The maximum rate of exploitation of the anchovetta reached ten million tonnes a year, equivalent to 6 mg of carbon for every square metre of the area. The upwelling in the Peru Current normally persists for nine months a year so that conditions are more constant than in the South Arabian coast region.

The upwelling on the South Arabian coast occurs only during the south-west monsoon for 4–6 months in a year. The whole ecosystem is adapted for this pulsing of the energy input. The main herbivores are zooplankton. The herbivorous zooplankton is eaten by small fish and carnivorous zooplankton. These first stage carnivores are in turn eaten by squid which are the main item in the diets of the tuna which are the main commercial species in the area. So between the primary producers, the phytoplankton, and the main commercial species are at least three intermediate stages. If the ecological efficiency of each of these stages is only 10%, any fishery off the South Arabian coast could not be expected to provide even 1% of the yield from the Peruvian Current region.

The amazing way in which the life histories of the individual species within a food web interlock is the simple food web associated with salmon larvae that has been described for some of the fiord-like inlets of British Columbia. The zooplankton is dominated by three species—a euphausid, *Euphausia pacifica*, and two copepods, *Calanus* and *Pseudocalanus*. In June and July there tend to be three successive blooms of phytoplankton. The initial bloom is of a silicoflagellate that does not seem to be exploited by the plankton. The next bloom is of the diatom, *Chaetoceros*, which is too large for the copepods to feed on but which is heavily grazed by the euphausids. The abundance of food stimulates the euphausids to spawn, as the *Chaetoceros* is grazed down. The diatoms are replaced by a bloom of $\mu$-flagellates only about 4 $\mu$m long. These are ideal food for the newly hatched euphausid larvae and the copepods, especially *Calanus*. Just as the $\mu$-flagellates start to bloom so the salmon fry start to feed on the euphausid larvae and the copepods. Thus the whole food web is delicately balanced to fit the succession of phytoplankton blooms and the spawning patterns and feeding of the plankton.

It is hardly surprising that biologists find it extremely difficult to predict what will happen to plankton populations when they are interfered with. Tropical food webs are much more complex than high latitude food webs and are much more fragile. Any interference with a stable tropical planktonic community is much more likely to cause unpredictable and irreversible changes than similar interference in temperate or polar communities.

Perhaps the biggest ecological experiment that has unwittingly been carried out by man is the removal of the whales from the Antarctic. These whales were the main predator of the krill and probably ate about 7,700 million kg of krill a day for the hundred days of the feeding season. Possible consequences are the faster growth and earlier maturation of the surviving whales and other krill predators such as the crabeater seals. There have been population outbursts of these and other seals and of several bird species, including the gentoo and king penguins. The Russians have now begun to exploit the krill directly. It would seem feasible that annual catches of krill could easily exceed the world annual catch of fish. It is ecologically more efficient for man to crop as close as possible to the primary producers, the phytoplankton. What is not clear is just what the ecological consequences would be. The whales may be saved from the whalers and yet threatened by the destruction of their ecosystem. And what other price will have to be paid for the exploitation of this small but important planktonic euphausid?

A diagrammatic representation of the food-chain associated with the tuna long-line fishery. Sunlight provides the energy for the growth of phytoplankton. These microscopic plants are consumed by planktonic herbivorous animals, which in turn are the food for carnivorous plankton such as arrow-worms, for oceanic birds including petrels and for various fishes including flying fish and lantern fish. Flying fish may be snatched in mid-flight by frigate birds whereas the lantern fish are attacked by squid. The squid are the staple diet of the tuna. With so many intermediate stages between the growth of the plants and the feeding of the tuna, the fishery cannot be expected to provide catches equal to even a hundredth of the original plant production.

# The Open Sea

Early studies of deep-living oceanic animals showed that an increasing variety of animal species are caught as the nets are towed deeper and deeper. However, the precision of the vertical structuring of the communities was not fully appreciated until samplers which could be opened and closed while fishing had been developed and the actual depth of fishing could be accurately monitored. There are five main zones of distribution of the midwater animals. Neustonic animals live right at, or on, the surface. The epipelagic animals inhabit the surface 250 m, the mesopelagic zone is at 250–1,000 m, the bathypelagic zone at 1,000–4,000 m and below 4,000 m is the abyssal zone. The depth of the transitions between zones varies from locality to locality depending on the transparency and productivity of the surface water, and, to some extent, on the temperature structure of the water column.

The inhabitants of each zone tend to have characteristic developments and modifications of their body structures which adapt them for better survival at the depths at which they live. These morphological adaptations are best known in the larger more powerfully swimming animals, usually called the nekton, such as the fishes, squids and prawns. The smaller planktonic species are distributed in identical zones to the nekton and probably possess similar, if less obvious, adaptations.

## Animals of the neuston

The neustonic species are strikingly different from the epipelagic species, which live only a few metres or even centimetres below them. But this specialized fauna associated with the surface layers occurs only in tropical and subtropical latitudes. Typically neustonic animals are blue and quite distinct from the transparent or colourless epipelagic species. The blue colour is either a protection against the strong ultraviolet radiation of tropical sunlight, or it is a camouflage against predation by birds from the air above, as the blue colour of the animals matches the background blue of the deep ocean. The blue pigment in many species is a complex formed between protein and a carotenoid. Carotenoids are a family of red pigments synthesized by plants and passed to a wide spectrum of animal species via their diets. By complexing the red carotenoid with the protein, its colour is modified to blue. The complex is unstable and neustonic animals kept in the laboratory soon show their lack of health by slowly turning reddish. The blue of the Portuguese man-of-war is, however,

HEATHER ANGEL

Goose barnacle larvae settle on all kinds of flotsam, from pieces of wood and seaweed to lumps of fuel oil.

Right: This flying-fish (*Exocetus volitans*) has powerfully developed pectoral fins which are clicked out to act as wings.

Pages 112–113
Squid are a numerous and very varied group of oceanic animals. This *Calliteuthis* species shows many of their characteristic features.

The deep ultramarine colour of these pontellid copepods is typical of the animals that inhabit the surface centimetre or so of tropical oceans.

DAVID/SEAPHOT

JACANA

produced by a compound related to bile pigments.

The ocean-strider, *Halobates*, is the only oceanic animal that lives its whole life above the water surface; it is a little insect related to pond-skaters. *Halobates* lays its eggs on any debris it encounters floating on the surface, from fuel oil and feathers to the bubble floats of the purple sea snail, *Janthina*. Fuel oil is also used as a base for settlement and metamorphosis by the cyprid larvae of the goose barnacles, *Lepas*. The oil seems to have absolutely no deleterious effect on the barnacles, which will even bend over and scrape up the surface of the oil and eat it; presumably they get nourishment from the bacteria degrading the oil. One species, *Lepas fascicularis*, secretes a bubble float from a gland at the base of its foot. The float then acts as a substrate for further larval settlement, so that colonies 5–10 cm in diameter are built up. Most *Lepas* species settle on floating pieces of wood or cuttle-bone. Such objects float for a relatively short time at the surface, so *Lepas* needs to grow very rapidly and can increase from its larval size of 2 mm to become a 2–3 cm long sexually mature adult releasing larvae in less than three weeks. Superficially the goose barnacles appear to lack any trace of blue pigmentation, but if the column is dissected open the gonads of these hermaphrodite animals are seen to be ensheathed with blue tissue.

Many other neustonic animals besides *Lepas fascicularis* have their own gas-filled floats which keep them at the surface. The gas in the floats of the Portuguese man-of-war (*Physalia*), the sailor-by-the-wind (*Velella*) and *Porpita* is curious in that it contains a large proportion of carbon monoxide, a gas which is highly toxic to animals like ourselves that use haemoglobin as an oxygen-transport system in the blood. In both *Velella* and *Physalia*, the float is offset so that they sail obliquely across the wind. *Physalia* can dip its float down into the water, behaviour which not only keeps the float moist, but also allows it to tack to and fro across windrows where fishing is better for this dangerous predator because prey is more concentrated.

## Flying animals

Many of the larger animals associated with the neuston have developed an aerial mode of escape. Flying fishes are an obvious and familiar example; their pectoral and pelvic fins can be clicked open into an aerofoil structure that enables them to glide tens of metres. The lower lobe of the tail fin is enlarged and the flying fish can keep airborne by giving itself an extra push with a few flicks of its tail which just dips into the water. Despite the apparent effectiveness of the flying fish's mode of escape, the dourado (dolphin fish) is able to follow the flight at high speed beneath the surface and grab and eat the flier when it eventually flops back into the sea. There is also a flying squid and the superbly ultramarine-coloured pontellid copepods which can leap well clear of the water. These little crustaceans have eyes specially adapted for their existence at the surface, with one part modified for seeing in water below the surface and the other for seeing up through the surface.

## Colour in epipelagic animals

The most familiar marine animals are those from the epipelagic zone, since the surface 250 m are the most accessible. Nearly all pelagic commercial fishes live in this zone. The pattern of colouring of familiar fishes like herring, mackerel and tuna is typical of fishes living in this zone by day. The back is dark blue, matching the colour of the deep ocean below. The belly is pale and silvery, giving a countershading effect which renders the fish less obvious when seen from one side. Also the flanks are often patterned with disruptive bars of darker colour which break up the fish's outline. Like nearly all general body patterning in oceanic animals, the colouration is a camouflage. Exceptions do occur, especially in reef fishes which are territorial and in which colour functions as an intra-specific display.

Warning colour occurs in the anemone fish and in a little fish, *Nomeus*, which lives between the tentacles of the Portuguese man-of-war and has conspicuous vertical dark blue and silvery banding. The pilot fishes which station themselves just ahead of large sharks, are similarly coloured to *Nomeus*, and this too may be a warning colouration. Mimicry is rather uncommon in oceanic fishes, but the *Sargassum* fish, *Histrio*, is almost indistinguishable from the *Sargassum* weed amongst which it lurks, waiting to gulp down any unfortunate fish that fails to notice its presence. In shallow fringing seas, mimicry is more common and, as we have seen, plays an important part in coral ecosystems.

Colour can only be effective for display when it can be seen. Animals living in the poorly lit or permanently dark depths of the deep ocean have camouflage colours and use bioluminescence or chemical signals instead of patterns for displays.

DAVID/SEAPHOT

In tropical seas, clumps of Sargassum weed floating at the surface create a special habitat, to which various animals have become specially adapted. The colour of the anemone and the trigger fish with its surface papillae, camouflage them against the weed.

Right: *Physetocaris* is a very rare and little known decapod crustacean from the central Atlantic. Its colouration, part transparent and part red, is typical of many of the crustacean shrimps and prawns which occur at depths of around 500 m.

*Nomeus* is a small fish that lives in amongst the tentacles of the Portuguese man-of-war. Normally it appears to be able to brush against the tentacles without triggering the lethal discharge of the stinging cells.

FAIN/NSP

## Colour in mesopelagic animals

At shallow mesopelagic daytime depths of 250–700 m, the planktonic species are more orange or reddish. Decapod prawns are half red and half transparent. Again the red colour is produced by carotenoid pigments and concentrated in regions that are difficult to render transparent such as the head, the gut and the gonads. At the depths at which these prawns live by day, there is no red light. The vast majority of these midwater animals are blind to red light, but, even if they were able to perceive it, the red parts of these prawns would still appear black as there is no red light to reflect.

The mesopelagic fishes are strikingly different from the epipelagic species. Their backs are black, tinted with a black pigment, melanin. The flanks are silvered like mirrors and the bellies are lined with light organs. In describing the function of the mirror sides, the reader needs to remember the symmetry in the distribution of light intensity below 250 m (Chapter 1). If a mirror is suspended in the water it will reflect light identical in intensity and colour to the background and so be quite invisible. The way in which the silvering is produced is a remarkable tribute to the precision of the structural architecture achieved by selective pressures. The tissues covering the fish's scales are packed with guanin crystals. Each crystal has a thickness of 0·25 ångstroms (1 Å = 1 mm/100 million), equal to a quarter of the wave length of the blue-green light that penetrates to the deepest depths in the ocean; the gap between the crystals is the same as their thickness. This arrangement of crystals produces an interference mirror; more familiar interference effects are the colours produced by thin oil films on water. At night the mirror sides are a potential liability as any flash of bioluminescence will act like a searchlight lighting up the fish. Most species must rely purely on their greater alertness and activity to escape, but *Valenciennellus* has black melanophore pigment cells which expand at night to draw a curtain over the mirror sides, making it virtually invisible to predators.

DAVID/SEAPHOT

## Ventral light organs and their functions

Since the brightest light comes from directly overhead, a fish will be starkly silhouetted if seen from directly below. Quite a number of fishes possess eyes which look vertically upwards, no doubt for looking for the silhouettes of their prey. Many midwater fishes have their bellies lined with light organs which help to break up the outline of their silhouettes making them less recognizable. The effectiveness of these batteries of ventral light organs (photophores) is enhanced if their light output matches the intensity of the down-coming daylight. In several species of lantern fishes (myctophids) there is a light organ inside the eye. It is thought that the nervous control of the light output of this photophore is linked directly with that of all the ventral photophores. So the light organ within the eye can be used as a standard, and its output matched with the intensity of the daylight coming from above.

In *Histioteuthis*, a common midwater squid, the undersides of the body and the tentacles are studded with photophores, all pointing at an angle of 45° to the general body axis. This corresponds to the normal body attitude of the squid in midwater, head-down at an angle of 45°. Two of the photophores, however, shine up into the animal's head. They shine through a transparent window on the underside of the head onto the lower half of a light sensitive ganglion at the back of the squid's brain. The other half of this ganglion is adpressed to the underside of another transparent window, this time on the top of the head. This elaborate arrangement is probably another intensity monitoring system.

In another squid, *Lycoteuthis*, the eye also is camouflaged with a row of light organs underneath it. Internal organs, such as the liver and ink sac, of transparent squid species are often either silvered or protected with light organs. There is a remarkable similarity in the ways in which two

DAVID/SEAPHOT

*Vinciguerria* is a typical shallow mesopelagic fish, living by day at depths of 300–600 m. The sides are silvery and mirror-like, the belly is studied with light organs. This specimen is parasitized by a copepod.

such different groups of animals as the fishes and squids have solved the problems of life in the sea. Such convergent evolution is the result of the very limited number of ways in which these problems of survival and reproduction can be solved.

## Vertical migration

Many of the shallow mesopelagic species migrate up into the epipelagic zone at night. The top of the deep mesopelagic zone (700–1,000 m) appears to be the lower limit for vertical migration by plankton but not for the nektonic species. A very few deep mesopelagic species migrate right up to the surface; one lantern fish has been recorded off the Canary Islands as swimming up from 1,250 m to the surface and back again each night. However, most daytime deep mesopelagic migrant species move up into the shallow mesopelagic zone at night, swimming downwards again at dawn. These migrations are not yet fully understood.

## Colour in deep mesopelagic animals

Once again there are distinctive colour differences between many of the deep and the shallow mesopelagic species. The decapod prawns are totally red. The red pigment is again a carotenoid which is concentrated in the cuticle of the shell or carapace. The fishes at these depths are uniformly drab black or bronze. Silvering is no longer advantageous since the intensity of daylight at these depths is so dim, only just detectable by the human eye, that any nearby bioluminescence will

The decapod prawn, *Notostomus*, shows the typical colouration of prawns living below 700 m. The scarlet colour is produced by a carotenoid pigment which is derived originally from phytoplankton in the surface layers of the ocean.

DAVID/SEAPHOT

light up a silvery fish at any time of the day. Ventral photophores still occur in many fishes, but they are much simpler structures, lacking the elaborate diaphragms, reflectors and lenses occurring in the light organs of many shallow mesopelagic species. This drab black colouration of fishes predominates down through the bathypelagic zone, with occasional exceptions, such as the whalefish, *Barbaroussia rufa*, which is red rather like the decapods. On the bottom, or in abyssal depths, an increasing number of fish are colourless. At these great, perpetually dark depths, even the animals themselves have given up producing light.

## Bioluminescence

Production of light is not entirely restricted to marine organisms, but there are extremely few marine groups that do not have species with the ability to produce light. Many species produce their luminescence directly by a chemical reaction, but in some the light is produced by luminescent bacteria that the animal cultures in special organs. The output of the light from the bacteria can be controlled mechanically by masking the organ either with a reflective layer or with black melanin pigment, so that the light is only emitted in the direction required through a filter whose transmittance can be varied. Alternatively, the oxygen supply to the bacteria may be regulated; without adequate oxygen they will not luminesce.

Virtually every function served by colour in well lit environments can be served by bioluminescence in the dimmer environments of the deep sea. The use of colour for camouflage and the function of breaking up the outlines of silhouettes has been seen in the ventral photophores of many mesopelagic fishes, decapods and squids. Luminescence can also be used as a distracting device against predators. Shallow-living squids and octopuses eject clouds of sepia ink, whereas deeper living species produce luminous clouds. The squid may either shoot away, leaving behind a coherent cloud, often much the same size and shape as itself, or produce a more diffuse cloud inside which it hides. A whole range of other animals, from searsid fish to decapod prawns and tiny planktonic ostracods, produce distraction clouds of luminescing secretions, produced from special glands.

Photophores are also used as head lamps. Round the islands of the East Indies occurs a fish, *Photoblepharon*, which has a large photophore containing luminescent bacteria just behind each eye. The

DAVID/SEAPHOT

In close-up the head of the deep-sea fish *Pachystomias* shows the large re-curved teeth, the huge eye and three light organs (photophores) on the cheek. Most light organs produce blue-green light, but the organ with the red reflector is unusual in emitting red light.

photophore can be rotated so that the light can be turned on and off. At night shoals of these fishes move up from depths of 50–100 m to feed in the shallows round the fringes of the steep-sided volcanic islands. Observers on the shore can see the shoals by the bright gleams of their photophores as the fish feed on the small planktonic animals. Several of the species of lantern fishes have big headlamp photophores on their snouts. Many of the mesopelagic stomiatoid fishes have very large cheek photophores just beneath the eyes. These photophores are under nervous control, the animal itself producing the light. The backs of the photophores are lined with a reflective layer, and the colour of light that the photophore emits is similar to colour reflected by the reflective layer.

Light organs may also be used to make an animal appear larger than it really is. The little amphipod crustacean, *Scina*, has light organs on the very extremities of its long antennae and on the tip of its abdomen. When its light organs flash, the amphipod, which may have a body length of 0·5–1 cm, must look two or three times its size.

In the myctophids (lantern fishes), the flanks are covered by patterns of photophores that vary from species to species. These patterns are ex-

tremely useful to the biologist with the task of identifying the fish and the fish no doubt use them in exactly the same way. It is extremely important for any animal in the deep sea to be able to recognize members of its own species, especially when they are reproducing. Interbreeding between species, even if the eggs are viable, usually results in infertile offspring, a result which benefits neither species. The possibility of interbreeding between closely related pairs of species is usually reduced by their mating activities being separated in time and/or space; in the ocean spatial separation can be horizontal or vertical. Even so, there are elaborate mechanisms whereby mistakes are avoided. There may be special chains of behavioural responses that have to be followed with complete accuracy before eggs and milt are released, very similar to the complicated sequence of actions and responses followed by mating sticklebacks in freshwater. So the pattern of light organs on the flanks of lantern fish are probably used to confirm its identity to another fish before mating occurs. The males have a special very large light organ on their backs close to the tail, which is possibly used in the courtship display.

A series of other observations, however, suggest another function for these big tail photophores in the male. In the Pacific, examination of the stomach contents of skipjack (a small species of tuna) showed that they contain mostly the males of one species of myctophid. Yet net catches of the myctophid consist almost entirely of females. The interpretation is that, at the approach of imminent danger, the males dart away with their tail photophores brightly flashing, attracting the attacks of the skipjack. The females in the shoal stay unobtrusively where they are and so, despite the males' sacrifice, get netted in large numbers by the fishermen.

The use of light organs as lures is widespread. The angler fishes have very elaborate lures or *escas*. The esca is often very similar in appearance to planktonic animals such as copepods. Its attractiveness is reinforced by its luminescence being emitted as sequences of flashes. The light is, in this case, produced by luminous bacteria which are kept enclosed in a blind sac. There is no evidence of the sac ever being open to the outside environment. The puzzle is just how the bacteria get into the sac. If they do not enter at some stage from the outside water, then they must be handed on through the eggs from parent to offspring. Another suggested function of the flashing lure of the female angler fish is to attract the male for fertilization.

DAVID/SEAPHOT

*Caulophryne jordani* is a deep-sea angler fish which lives below 1,000 m. The head and the lure are covered with fine filaments which are rich in sense cells. In many species the lure is bioluminescent to attract the prey.

In the stomiatoid fishes the chin barbel is elaborated into a lure. Any movement close to the lure unleashes a frenzy of snapping and biting, so it must be covered with sense organs to detect the approach of the prey. The lure is so important to the fish that any damage to it must threaten its survival. So the prey must be grabbed and eaten before it actually swallows the bait. The viper-fish, *Chauliodus*, overcomes this problem by having light organs inside its mouth. The lure in this case is the first modified fin-ray of the dorsal fin. The prey is enticed on first by the lure and then by the light organs inside the mouth.

The function of bioluminescence is not always known. For example, many of the bottom-living animals at depths of 500–2,000 m have quite elaborate light organ arrays which seem quite pointless. Many of the ophiuroids (brittle-stars), have long slender arms which are picked out in points of light. Since they have no eyes, the light organs cannot be signals to members of their own species. Similarly, since they are mostly detritus and filter feeders, they cannot be attracting their prey. The light organ would seem to be more likely to draw the attention of the brittle-star's predators and so be a liability. The parchment-tube worm, *Chaetopterus*, is another curious example, since the worm lives permanently buried in mud enclosed in its tube at or just below the low water mark. Despite spending its entire life secluded and out of sight, this worm is brightly luminescent.

## Adaptations to life at depth

Bioluminescence is one of the most outstanding features of the oceanic fauna, but there are many other adaptations to life at depth. One of the most important is associated with the ever increasing scarcity of food with increasing depth. The tuna is a typical epipelagic fish living at depths where food is abundant. It is a highly streamlined fish, able to swim powerfully and continuously. As it swims constantly it has no need of a buoyancy organ and so has no swim-bladder. The eyes are well developed but are not unusually large. It hunts by sight so the lateral-line system is not elaborate. The most interesting feature is the wedge of red muscle that runs the length of the body. This special muscle is rich in haemoglobin and has a special counterflow blood supply. The counterflow system is arranged so that the incoming blood supply flows close alongside the outgoing blood supply, and this acts as a heat exchange system. The muscle generates heat by its continuous activity. This heat is retained within the muscle, maintaining its temperature as much as 10°C above that of the surrounding water. The red muscle can generate three times the power of the equivalent amount of white muscle, and gives the fish great stamina. Thus tuna and many other of the fast swimming pelagic fishes and sharks are, to some extent, warm-blooded.

*Chauliodus*, the viper-fish, is a typical example of a shallow mesopelagic fish with silvery sides, ventral photophores and a lure. An immediately obvious feature is the huge mouth and long recurved teeth. Deep-living fish have to be much less choosy about the size range of food they are prepared to eat. Many are able to eat fish and

The lure on the end of the barbel beneath the chin of *Astronesthes gemmifer* is luminescent. There are ventral photophores along the belly and a cheek light organ below the eye. The large re-curved teeth and the bronzy colouration are typical of deep-sea fishes.

DAVID/SEAPHOT

The impressive array of teeth of *Pseudoscopelus* ensure that once a prey animal begins to be swallowed it will never escape. At the depths at which this fish lives, food is in short supply and nothing must be allowed to get away.

DAVID/SEAPHOT

prawns as big as themselves. Not only do they have enormous gapes (*Chauliodus* can dislocate its jaw to extend its gape), but their stomachs need to be enormously extensible. The bathypelagic fishes, *Saccopharynx* and *Eurypharynx*, have gone to such an extreme that they seem to be little more than huge mouths opening into enormous stomachs.

*Chauliodus* has comparatively much larger eyes than the tuna, the larger lens and wider pupil being needed at dimly lit depths to collect the little light available. Tubular eyes are a common feature in these mesopelagic fishes. The hatchet fishes, for example, have tubular eyes that look upwards.

Fishes of the genus *Scopelarchus* also have upwardly directed tubular eyes with the retina divided into three distinctive areas, each area having different arrangements of the rods (light-sensitive cells). The walls of the tube are lined with an unusual type of retina that can be only light sensitive and unable to form an image. Light falls on this area, not via the lens but through a clear window in the side of the eye that 'looks out' sideways. Even the normal retinas of mesopelagic fish like *Chauliodus* tend to have two or more layers of rods, compared with the single layer in the eyes of epipelagic fish. The rods in the eye of *Chauliodus* contain nearly ten times more visual pigment per unit area and so may be 15–30 times more sensitive to dim light than the rods in the human eye. *Chauliodus* does not possess red muscle and so must be a much more sluggish fish. It either lures its prey gradually towards it, or slowly stalks it, grabbing it after a final quick dart. Such a fish must have a buoyancy system, so that it does not have to swim continuously to maintain its position in midwater. Like many midwater fish, *Chauliodus* has a gas-filled swim-bladder. These swim-bladders reflect the sound used by echo-sounders to chart the depths of the ocean, and concentrations of fish appear as a false bottom on the echo-sounder record. These false bottoms are called deep scattering layers. The special gas in the swim-bladder is derived from the atmospheric gases dissolved in the sea-water and is secreted into the bladder through special retia cells lining a part of the swim-bladder wall called the *rete mirabile*. The retia cells have to work against the hydrostatic pressure, so the deeper a fish lives, the more efficient and progressively more expensive in energy the retia cells have to be. Swim-bladders tend to be degenerate in fishes living at depths below about 700 m, and alternative systems of maintaining buoyancy are developed.

One such alternative is to increase the oil and fat content of the body; both of these are substances which are much lighter than water. This is often accompanied by weight-saving economies, such as the reduction in the degree of calcification of the skeleton and in the quantities of body musculature. Thus comparing two species of fishes of the same genus which live at different depths, the shallower living species will tend to be deep-bodied with a functional swim-bladder and a heavily calcified skeleton; the deepest living species will be the slimmest and have a regressed swim-bladder and the most weakly calcified skeleton.

Another widespread mechanism amongst animals is the selective exclusion of heavy sulphate ions from the body fluids. This method of keeping the body at the same density as the surrounding sea-water is most frequently found amongst gelatinous animals like jellyfish, but it also accounts for the tendency for gigantism in some deep living animals. It occurs in the world's largest ostracod, *Gigantocypris*, and in some of the large gelatinous cranchid squids and the midwater octopus, *Japetella*.

Animals invading deeper and deeper depths are faced with two conflicting pressures. One is a need for progressively more elaborate sensory mechanisms to locate and capture the food, because of the decreasing availability of food with increasing depth. The more elaborate a sensory mechanism the more expensive it is to run so that there is ever increasing diminution in the return for the energy expended. The second pressure is towards simpler, cheaper-to-run, body systems. Thus the numbers of actively predacious animals decrease with depth because of the increasing danger of an animal exhausting its food reserves before it finds and catches its next meal. The big switch seems to occur at about 1,000 m, the maximum depth to which daylight penetrates, and the depth at which the availability of living organisms or biomass is less than a thousandth (or three orders of magnitude less) than in the photic zone.

Eyes of many of the species regress to such an extent that effectively they are blind. Thus many bathypelagic species that have active larval stages in the near surface layers, lose some of their sensory systems as they metamorphose into the adult form. The pace of life is slowed, they live a more passive existence, patiently waiting for the eddies and swirls of the deep water to drift them, by chance, towards potential prey. They then slowly stalk the prey using their senses of taste and the lateral line system. Fishes such as the

OXFORD SCIENTIFIC FILMS

The hatchet fish (*Argyropelecus hemigymmus*) occurs at depths of 300–400 m. Its flanks are mirror-like and the belly photophores contain a combination of fully and half mirrored walls. The tubular eyes look vertically upwards, searching for the silouhettes of its prey against the dim light above.

angler fishes probably do not even stalk their prey but use their lures to entice it closer. They possess an elaborate lateral line system that must allow them not only to sense the prey's approach but also to accurately pin-point its position. The mouths of the anglers are immense for their size, so they can swallow and slowly digest other fish as large as themselves. Life in the deep ocean is very much a matter of eat or be eaten.

## Reproductive adaptations

Besides feeding, the need to breed successfully is a basic requirement for a species' survival. Many deep bathypelagic species are extremely rare and the volume of water through which an individual needs to search for a mate is immense. We understand very few of the mechanisms by which animals in breeding condition find each other. As far as the commercial fishes are concerned, fishermen, over the years, have learnt the areas in which breeding shoals accumulate in vast numbers thus making catching the fish a commercially viable proposition. Out of the breeding season, the fishermen either switch to exploiting other stocks or put up with much reduced catches. Thus the herring in the North Sea has about six main breeding sites, each occupied by a different stock of fish, each breeding at a slightly different time of year. During breeding the herring shoals may be concentrated into a tiny area only a few hundred metres across. It is hardly surprising that many of these stocks have been fished almost out of existence.

The deep sea fishes are never able to muster a population even one millionth of the concentration achieved by the pelagic fishes. The search for a mate is usually carried out by the male which is sometimes specially adapted for the search; an extreme example is some of the angler fishes. The female is a relatively big lumbering fish, adapted to a life of patiently luring prey towards herself, only breaking into a frenzy of activity as food comes within catching distance. Her eyes are small, her sense of chemical perception or 'smell' is weak, and her muscles are all white and flabby so she is unable to sustain prolonged activity. The male in contrast is a tiny fish probably less than a quarter of the female's length. He has enormously developed eyes and huge nasal rosettes with which to smell out the female's secretions. He is unexpectedly well endowed with the red type of muscle which, as we saw in the tuna, is an adaptation to prolonged and sustained swimming. Thus the male metamorphoses from the larval form at shallow depths and swims down into the depths to find a female. He probably only has enough food reserves to last a few days. The female is probably initially sensed by taste and then homed in on through the species-specific pattern of flashing by her bioluminescent lure. His speed and smallness probably enable him to dodge her lunges, as to her he appears to be just a potential lump of food. He attaches himself to her flank and then lives as a parasite on her, fulfilling his function by fertilizing her eggs as they are shed.

Another reproductive adaptation to life in the deep sea is that of hermaphroditism. This may take the form of a fish maturing as a functional male and then undergoing a sex reversal to become a functional female. This was first recorded in *Gonostoma gracile* from the Sea of Japan, but other examples have recently been discovered. In the rare mesopelagic fish, *Benthabella*, each mature fish is a complete hermaphrodite able to function simultaneously as a male and female. There is no obvious anatomical barrier to self-fertilization; a possible solution of how to be extremely rare yet still able to reproduce successfully.

## Life on the sea-bed

Life on the bottom must be a little easier than life in midwater. The bottom is a surface that automatically collects particles and organic debris settling out of the water above. There is a choice of life styles open to the organisms; they can filter-feed the fine particles carried past them by the water currents. They can burrow within the sediments, ingesting any organic matter the muds contain. They can feed carnivorously on the other animals or scavenge on the corpses of the dead giants of the sea, such as whales and large tuna.

In the shallow fringing seas, organic debris such as terrestrial plant remains, pieces of seaweed and insects blown out to sea, all provide a rich source of food for the bottom-living or benthic animals.

DAVID/SEAPHOT

This larva of the fish *Xenolepidicthys* looks as though it is swimming at an angle of 90° to its real orientation. Presumably predators aim just in front of their prey when they strike and so will tend to miss.

On a cliff-face at a depth of 35 m in Antarctica the rock is covered by encrusting animals in amongst the seaweeds. Most are filter-feeders, extracting nourishment from the rich supply of fine particles suspended in the water.

SHABICA

ARIES II MISSION

The rich abundance of Antarctic life occurs even at a depth of 3,155 m. Here many deep-sea fishes, rat-tails and brotulids have been attracted to the bait attached to the arm of a free-fall camera.

*Monognathus* is amongst the strangest of all deep-sea fishes. Coming from a great depth where there is no light by which to see, or to be seen, it is blind and colourless. The stomach is greatly swollen with its last meal of planktonic animals.

DAVID/SEAPHOT

Farther out to sea this rain of organic material from the land diminishes and dies away. The amount of food reaching the bottom then reflects the richness of the surface water above. Like all animal life on our planet, the benthic animals are ultimately dependent on photosynthetic activity of the plants growing in the sunlight, but how this food reaches the deep community is still a subject for debate.

Round the edges of the continental slope, turbidity currents may play an important role in the rapid transport of organic material into the deep ocean. These are great boluses of water made heavy by large quantities of suspended muds, which cascade down the slope into very deep water. They may erode away the shelf edge cutting back great canyons, some of which are of much grander and majestic dimensions than the Grand Canyon of Colorado.

Away from the continental slopes, it is generally considered that the rain of fine debris from the photic zone is the main source of food supply for the bottom living animals. Small particles, however, sink so slowly that they are degraded by bacterial action well before they sink more than a few hundred metres. Measurements of the particulate organic material in suspension in sea-water show that it decreases greatly with depth and is almost negligible below 1,000 m. Thus the potential food must be accelerated to the bottom in some way. One way is via the migration of the midwater animals. Nekton may feed near the surface at night and migrate down to 1,000 m or so and there excrete the undigested part of its food. Certainly sediment traps set in deep water have shown that faecal pellets, even from non-migratory near-surface-living animals like salps, sink rapidly enough into deep water to be an important food source. There is also evidence of bottom fish living at depths of 1,500–1,000 m migrating up into mid-water and feeding on mesopelagic fishes, but this activity is probably restricted to the edge of the continental shelves where there are large midwater fish populations.

Films taken with deep-water cameras show that, even in the deepest depths the bait rapidly attracts a large number of scavenging and predatory fishes. These remarkable pictures show the way in which the fish swim very slowly, almost as if they were swimming through treacle. Yet the speed at which they gather round the baits indicates that these fishes are unexpectedly abundant, giving credence to an idea first put forward in the 1930s by Krogh. Krogh pointed out that large fish and whales sink when they are killed and, since they are too big for most midwater scavengers to make much impression on, would rapidly reach the bottom. Thus very large packages of food are likely to arrive on the bottom at irregular intervals, but possibly frequently enough to nourish these big populations of bottom-living scavengers. The scavengers, in eating the dead carcase, would fragment much of it into suspension in the water and, since they are far less than 100 % efficient in ingesting all they eat, still more would be dispersed widely to the more sessile animals through their faeces.

Bottom-living fishes are quite distinctive in their shape and structure from midwater fishes. Most of those living below 200 m tend to be colourless; the black sharks are an obvious exception. Species living to 2,000 m are mostly substantially built fish with well developed sensory systems, large eyes, large hearts, large brains and many possess large swim-bladders. Bottom fish tend not to use the swim-bladders primarily as buoyancy organs, but more for sound production and reception. Male rat-tail fish (macrourids) have drumming muscles attached to their swim-bladders whereas the females do not. Both sexes have large otoliths, which are calcareous structures used to listen to sound waves. The mating call of the rat-tail fish is a series of clicks, whereas in the deep-sea cods the mating call is a grating noise produced by grinding their pharyngeal teeth.

A feature very common in bottom-dwelling fish in deep water is a long filamentous tail. This is associated with the elongation of the lateral line system, which is sensitive to low frequency vibrations and movements in the water. The longer the array of such receptors, the more accurately can the source be pin-pointed and the speed and direction of the movement assessed. In active midwater fish, the sensory receptors of the lateral line system, the neuromast cells, are enclosed in a mucus-filled canal, but in many bathypelagic fishes they are free-ending to the surrounding water. The benthic fishes have a mixed system of some free neuromasts and some in canals.

The tripod fishes (bathypteroids) have a fin-ray greatly elongated in each of the pectoral fins and the lower lobe of the tail (caudal fin). They prop themselves up above the bottom on these long fin-rays as if they were stilts, using their well developed sense of taste or 'smell' to scan the water for any hint of food carried in the current.

Below 2,000 m the bottom-living fishes tend to be built for economy. The eyes especially are often degenerate; this is of value to the biologist since the fishes seem totally insensitive to the lights needed for taking ciné-film and carry on their normal behaviour quite unconcernedly. Even so many of these fishes have well developed tracts of red fibres in their body musculature suggesting that they actively search out their food rather than lie in wait for it.

The sea-bed is also thought to be the home of the big squids that sperm whales feed on. Plotting the distribution of occurrence of sperm whales shows that they are often hunting in water of up to 2,000 m depth and from the size of giant squids found in their stomachs, it is clear that they will attack and eat even the largest. (See also Chapter 7.)

## Feeding adaptations

The squids, like many of the fishes, are carnivores, or possibly scavengers. The mode of feeding of squids cannot be assessed from their shape and structure, whereas, in the fishes, those that feed above the bottom have their mouths on the front of their heads. The fishes that predate animals living in and on the sea-bed tend to have their mouths beneath their heads. Their feeding habits have been confirmed by examining the contents of their intestines; in the latter group of fishes, the intestines contain large quantities of sand and mud; in the other group they contain only animal remains. The prey of the bottom-feeders include a wide spectrum of animal groups, most of which belong to shallow water and shore species, although the actual specific composition is quite distinct.

In regions where the sedimentation rates are extremely high and water currents are strong, filter-feeding forms predominate. Their body shape and structure may change quite markedly, depending on the usual strength of the currents they encounter. The Devonshire cup-coral, *Caryophyllia smithii*, was originally thought to be two species, one with a very broad base that occurs in areas of high currents, the other with a narrow base that grows in areas of low current. Where the currents are strong, many of the filter-feeders are either massive or very flexible so that they bend with the current. Sea-pens and sea-whips, are typical of these flexible forms. Where the currents are gentle the filter-feeders have much more elaborate and delicate bodies. Some sea-fans and glass sponges have exquisitely patterned body structures with the optimum design to expose the maximum filtration surface to what current there is. Sometimes the shape of the bottom creates microhabitats in which animals can survive and live successfully in unexpected areas. For example, off the west coast of Scotland, geologists have identified scour marks in the sea-bed on the edge of the continental shelf. These turn out to be plough marks gouged by icebergs during the last ice age. Within these grooves in the bottom, protected from the full strength of the scouring tidal currents, grow extensive colonies of the cold water coral, *Dendrophyllia*.

The behaviour of animals may change with the strength of the currents. Brittle-stars (ophiuroids) are abundant representatives of the bottom fauna. Some of the species are purely carnivorous, eating newly settling larvae or some of the more microscopic bottom animals, such as the Foraminifera. Others can alternate between being carnivores or filter-feeders, depending on the availability of the food supply.

Where the sedimentation rate is slower but the bottom muds are rich in organic matter, detritus-feeders become more dominant in the bottom communities. In areas where the rates are higher, their relative abundance is lower. These are animals such as sea-cucumbers, worms, clams and snails which feed by swallowing the mud, getting their food from the organic matter it contains. The highest concentration of organic matter usually occurs in the surface millimetre of the mud, so many of the animals suck up this superficial, often very liquid, layer of mud, rather like a vacuum cleaner. Elasiopod holothurians crawl over the mud surface sucking up this surface layer; some of the sea-urchins (echinoids) just scratch up this layer, and some of the clams have long siphons which they extend out to sweep clean the sand surface in an arc around them. However many other species live an earthworm-like existence, only in a marine context, burrowing through the sediments. Perhaps the most bizarre animals are the Pogonophora. These are little worm-like animals which live totally enclosed in thread-like tubes and have neither a mouth nor a gut. How these animals feed is something of a puzzle, but recent evidence suggests that they can absorb and live off the minute traces of organic compounds that are dissolved in the interstitial water of richly organic muds. Even in these areas of moderate sedimentation, filter- and suspension-feeders can locally become predominant, wherever the currents sweep the sediments clear of the rocks, so providing them with a firm base on which to settle. These are favourite areas for sea-squirts, sea-lilies and sponges.

In very deep water, sedimentation is very slow and all the readily utilized organic material has been degraded by bacteria well before the sediments reach the sea-bed. Below the carbonate compensation depth the bottom deposits consist mainly of red clays with sharks' teeth and manganese nodules forming the solid surfaces for sessile animals to settle on. The muds are too poor to support detritus-feeders and once again suspension-feeders become predominant, but the weight of animal per square metre may be less than 0·1 g. Suspension-feeders are able to subsist because their food-collecting systems require very little energy, especially if they utilize the slow drift of water to carry their food to them. Their growth rates must be exceedingly slow and it has been estimated that some of the tiny 3–5 mm long clams that occur at great depths may be over a century old.

## Animals in the abyss

Despite its paucity the ultra-deep sea fauna is rich in species and extends into the greatest depths sampled. Foraminifera, sea-cucumbers, crustaceans (including isopods and amphipods) and snails have been brought up from depths of 10,000 m. The deepest a fish has been caught is 7,160 m, although pop-up cameras have photographed fish at greater depths. The fauna is almost entirely unknown; a recent Russian expedition sampling in the deep trenches round the West Indies collected large numbers of isopod crustaceans (relatives of the woodlice in the garden) of which three-quarters of the species were totally new to science.

## Living fossils

The deepest parts of the ocean are not likely to support unknown sea monsters because of the paucity of the available food. Even so, occasionally biologists find themselves faced with an animal that looks familiar because of its similarity to fossil forms. A little snail called *Neopilina* was recently discovered in deep sea dredge hauls from the trench off the Pacific coast of Costa Rica. The closest relatives of this little snail are found in the rocks laid down in the Cambrian era over 500 million years ago. More famous was the discovery of the coelacanth off the east coast of South Africa. This is an interesting fish since, along with the lungfish, it shows the beginnings of the evolution of fins into limbs and some of the prerequisites for the evolution of a terrestrial existence. The name given to this living species of coelacanth was *Latimeria* and quite a number of these fishes have now been caught, mostly off the Comoro Islands which lie between Africa and Madagascar. Any study of the fauna of the deep sea is always tinged with the excitement of the possibility of discovering new animals which will give added insights into our understanding of the geological history of life on Earth.

# The Air Breathers

Blue whale *Balaenoptera musculus*
mean length 24m, mean weight 84 tonnes

Fin whale *Balaenoptera physalus*
mean length 20m, mean weight 50 tonnes

Grey whale *Eschrichtius robustus*
mean length 14m, weight 32 tonnes

Humpback whale *Megaptera novaeangliae*
mean length 13m, mean weight 33 tonnes

Sei whale *Balaenoptera borealis*
mean length 16m, mean weight 23 tonnes

Minke whale *Balaenoptera acutorostrata*
mean length 9m, mean weight 7 tonnes

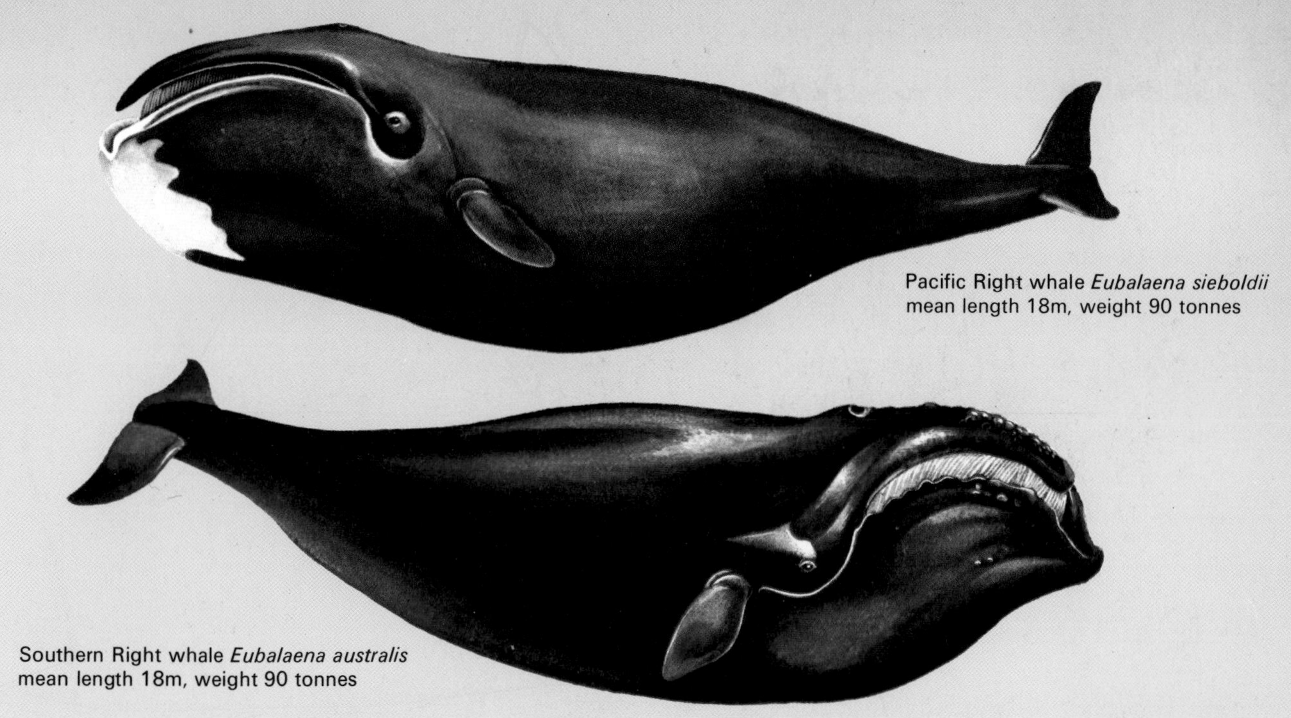

Pacific Right whale *Eubalaena sieboldii*
mean length 18m, weight 90 tonnes

Southern Right whale *Eubalaena australis*
mean length 18m, weight 90 tonnes

The eighty living species of whales include also dolphins and porpoises. They fall into two distinct groups, whalebone or baleen whales and toothed whales. The whalebone whales (left and above) are large and tend to stay close to the surface of the water. Their 'whalebone' is sheets of springy, fibrous material with frayed edges, through which planktonic animals are filtered from the sea. The toothed whales (below) are generally smaller—only sperm whales match the baleen whales in size—but take bigger prey, including fish, squid and even seals. Killer whales especially have a reputation for ferocious hunting. Sperm and bottlenosed whales, especially, dive deep for their food, but all whales must come up to the surface to breathe.

Bottlenosed whale
*Hyperoodon ampullatus*
mean length 9m, weight 8 tonnes

Killer whale *Orcinus orca*
males up to 8–9m, females 7–8m, weight 10 tonnes

Southern Bottlenosed whale
*Hyperoodon planifrons*
mean length 10m,
weight 8 tonnes

Dusky dolphin
*Lagenorhynchus obscurus*
mean length 2m, weight 200 kg

Narwhal *Monodon monoceros*
mean length 4·5m, weight 1·5 tonnes

Beaked whale *Berardius arnuxii*
mean length approx 10m, weight 7·5 tonnes

Cruciger dolphin
*Lagenorhynchus cruciger*
mean length 2m, weight 100kg

Beluga whale *Delphinapterus leucas*
mean length 5m, weight 2·5 tonnes

Blackfish *Globicephala* sp.
males average 6m, females slightly smaller,
weight 3 tonnes

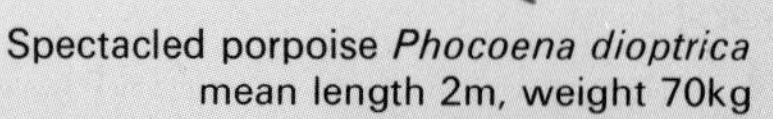

Spectacled porpoise *Phocoena dioptrica*
mean length 2m, weight 70kg

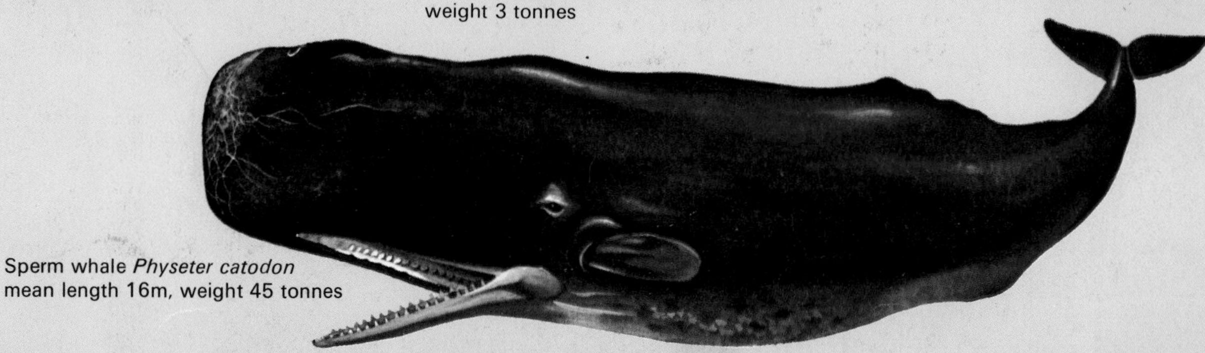

Sperm whale *Physeter catodon*
mean length 16m, weight 45 tonnes

The oceans offer an environment in which a range of air-breathing vertebrates live, find food and even breed. Every successful terrestrial group of vertebrates includes a few maritime species, with the exception of the amphibians. Amongst the reptiles there are sea-snakes, turtles and marine iguanas. There are many oceanic birds that feed by diving underwater, while sitting on the surface or by snatching food from the water as they fly over it. All the birds, and most of the reptiles, return to land to breed. Similarly, many of the marine mammals haul themselves out on land or ice-floes to breed. However, two groups of mammals have become totally aquatic; these are the cetaceans—the whales and dolphins, and the sirenians—the dugongs and manatees, now in danger of extinction.

## Reptiles of the oceans

In the age of the dinosaurs, which ended some 64 million years ago, there were many oceanic reptiles. The big fish-eating pleisiosaurs and ichthyosaurs must have been superseded by the more successful toothed whales. The marine reptiles are a motley collection of oddities. The four species of marine turtles are the most fully adapted to life in the sea. The largest of these is the leathery turtle, *Dermochelys coriacea*, which often exceeds 2 m in length and 500 kg in weight. They feed on jellyfish and salps, but little is known of their habits at sea: they probably frequent deeper waters than other turtles and there they are often accompanied by pilot fish.

The most heavily exploited turtle is the green turtle, *Chelonia mydas*, so named because of the

Right: The loggerhead turtle (*Caretta caretta*) is an omnivore, eating fish, jellyfish and seaweed. Whereas the flesh of the other turtle species is good to eat, the loggerhead's is tough and stringy, so it is seldom hunted for food.

The hawksbill (*Eretmochelys imbricata*) is one of the turtle species that is still abundant although before the invention of plastic its shell was eagerly sought to make tortoiseshell combs. Barnacles and swimming crabs often occur living on its carapace.

HUTSON/NSP

Left, below: A female green turtle (*Chelonia mydas*) laboriously hauls herself up a beach in search of a place to lay her eggs. In her element, the sea, she is a graceful swimmer, but on land she can hardly move.

Sea-snakes like this banded sea-snake (*Laticauda laticaudata*) have flattened tails for swimming. Although their bites are often fatal, they rarely kill humans because they are back-fanged.

NEVILLE COLEMAN

BANKS/NSP

TAYLOR/ARDEA

green tinge of its fat. Both the eggs and the meat are good eating, possibly because the usual diet consists of turtle grasses instead of the fish, crustaceans and jellyfish which the other turtles eat. Green turtles make very extensive migrations. Marked individuals have been seen over 2,000 km from their 'home' beaches.

The hawksbill turtle, *Eretmochelys imbricata*, was exploited for another reason. Until the invention of plastic, the carapaces of these turtles were used for tortoise-shell objects such as combs.

The fourth marine turtle species, the loggerhead turtle, *Caretta caretta*, is an omnivore, eating fishes, molluscs, jellyfish and some seaweeds. Since the meat is tough and stringy, loggerheads are not killed for food in anything like the numbers of the green turtles.

In some ways the fifty species of sea-snakes are more completely marine than the turtles. Many of the species are ovoviviparous, so that the young are live born; a few species, however, still need to come ashore to lay their eggs. Sea-snakes frequent estuaries, sandy coasts and coral reefs in the tropical parts of the Indian Ocean and throughout the East Indies. An isolated group occurs in the Gulf of Panama, in the warmer Atlantic waters.

The colouration of many species is counter-shaded, i.e. dark on the back and pale on the belly. In contrast, species that frequent seaweed beds have alternating rings of black and grey-green along the length of their bodies. On the very tip of the head are the nostrils, which are valved. Sea-snakes are able to stay submerged for up to eight hours. They feed mostly on small fish. Most sea-snakes reach only 1·5 m in length and the longest species grows to only 3 m, yet these little snakes are some of the most venomous known. Luckily they are back-fanged, so few people have been lethally bitten, but there is no known antidote for the bite of many species. At times, sea-snakes mass in huge numbers. They occur in dense concentrations during storms in mangrove swamps. The most remarkable record was of snakes, swimming at the surface in the Malacca Strait, in a belt 3 m wide and nearly 100 km long!

One of the more dangerous marine reptiles is the sea-water crocodile, *Crocodylus porosus*. It frequents the shores of south-east Asia and the north coast of Australia. Growing to lengths of up to 6 m, this fish-eater has been known to attack men when it is provoked.

An unsuspecting visitor to the Galapagos Islands

HEATHER ANGEL

Islands are often inhabited by unique endemic species. Colonies of the marine iguana (*Amblyrhynchus cristatus*) occur only on the Galapagos Islands. These seaweed eating lizards are able to swim and dive to depths of over 30 m.

might be horrified by the grotesque hordes of marine iguanas that, in certain places, cover the laval rocks in their thousands. Yet these placid animals are quite harmless and feed solely on seaweeds. They spend most of the early morning basking in the sun. Late in the afternoon they move down to the water and start to feed. They can dive quite happily to depths of over 20 m.

## Sea-birds

Birds exploit the whole of the ocean surface for food. There is, however, a direct correlation between the abundance of the birds and the richness of the surface plankton. There are three main zones to which each group of birds restricts its feeding. The *inshore zone* extends from the shore line to a few kilometres offshore. Birds that feed in this zone include many gulls, cormorants, terns, waders, brown pelicans and the few maritime raptorial birds. Many of these birds have continental distributions which are interrupted by oceanic barriers. The *offshore zone* extends a little way beyond the shelf edge, but can be very extensive where the continental shelf is very broad. It is a zone occupied by gannets, tropic birds and auks.

The *pelagic zone* covers much of the open ocean and this is exploited by the truly oceanic birds—shearwaters, storm petrels and albatrosses. These are birds which, in contrast to the inshore feeders, have trans-oceanic distributions which are interrupted by land barriers.

The oceans offer a predictable food supply to the highly mobile birds that can seek out the patches of high productivity. There is an absence of avian predators (but not pirates) amongst the oceanic birds. It is only on their breeding grounds that many of these species experience predation pressures. Possibly as a consequence of this 'secure' life, many species lay only single eggs. Petrels, tropic birds, frigate birds, gannets, auks and most penguins, rarely lay another egg if the first is lost. Another reason for laying single eggs is the problem

Gulls abound in the rich coastal waters around the continents. Here a flock of herring (*Larus argentatus*) and lesser black-backed gulls (*L. fuscus*) have been attracted to feed on fish and offal. These raucous scavengers rarely feed out of sight of land.

Penguins are flightless birds whose wings are adapted for swimming. They are restricted to the southern hemisphere and apart from the Galapagos species, they all occur south of the tropics. This is a black-footed penguin (*Spheniscus demersus*) from South Africa.

of rearing a hungry chick which may be waiting on an island as much as a hundred kilometres away from the feeding grounds.

Many of the pelagic feeders have long life expectancies and also have extended adolescences. King penguins and fulmars take five years to reach sexual maturity and the Royal albatross takes nine or more years. Since most of these pelagic feeders come ashore only to breed, there are always large numbers of non-breeders out at sea throughout the year. In addition, many breed only in alternate years. This is particularly common in some of the penguins that have an extended fasting period while the chick is brooded and reared.

**Penguins** Penguins are the most highly adapted marine birds. They have totally lost the ability to fly. Instead, the wings are modified into narrow flippers which are the sole mode of propulsion underwater. They are restricted to the southern hemisphere, and appear to have been so always; their first fossil records are found in early Tertiary deposits about 50 million years old in Antarctica. They are also restricted to the colder currents: several species occurring off New Zealand, and others off Australia, South Africa and South America. One species, the smallest of all, occurs on the equator, in the regions where cold water upwells around the Galapagos Islands.

The faster species can swim at speeds of over 18 km per hour. Adélie penguins 'porpoise' out of the water as they swim, and can leap on to ice-floes four times their own height above water. On land, some species can hardly hop or waddle; in contrast an Adélie can outstrip a man running. Many of the Antarctic species readily flop on their bellies and toboggan down ice-slopes.

The largest of all penguins is the Emperor penguin, *Aptenodytes forsteri*. Standing over a metre high and weighing about 40 kg, an Emperor penguin is the heaviest oceanic bird. It also holds the record for deep diving; a depth gauge attached to one specimen recorded a dive to 265 m. At such depths they hunt squids and fishes, whereas the chinstraps and gentoos, which are the smaller Antarctic species, feed mostly on krill. All penguins are extremely gregarious during breeding, and they congregate in extensive rookeries. Penguins occurring further north in the sub-Antarctic waters round New Zealand, South Africa and South America, nest in burrows and hollows. Their breeding sites are close to food sources and so feeding of the chicks follows a normal avian pattern. Most southern penguins must fast while brooding.

**Auks** It is interesting to note that the original penguin, the great auk, *Pinguinus impennis*, occurred in the North Atlantic. This large flightless auk was exterminated by fishermen taking large numbers of both birds and eggs for food. Auks are a northern hemisphere group of sea-birds. They swim underwater using their wings, steering with their legs, and all the living species have the ability to fly. However, most of the 22 species moult all their flight feathers simultaneously, and during this moult they are flightless. Their food consists mainly of fish and crustaceans caught in midwater, although a few species feed near the bottom. The little auk, *Plautus alle*, has the remarkable ability to fly into a wave and straight out the other side, apparently snatching its food on the way through.

**The tube noses** This family of birds includes the albatrosses, fulmars, shearwaters and petrels. It is predominantly a southern hemisphere group which probably evolved in the rich, windswept expanses of the Southern Ocean. In the 'Roaring Forties', the 'Furious Fifties' and the 'Screaming Sixties', mastery of the gliding flight makes circumnavigation of the earth almost effortless. It is in the latitudes of the West Wind Drift that the albatrosses evolved. They exploit the uplift on the windward side of waves to glide and wheel continually, rarely needing to flap their long narrow wings. Albatrosses are almost unable to take off from the ground, some need to jump off cliffs into the wind to get airborne, just like human beings hang-gliding. Each species has only a few breeding sites. The largest of all, the wandering albatross, *Diomedea exulans*, nests on the sub-Antarctic islands of South Georgia, Tristan da Cunha, Kerguelen and Auckland.

The medium sized tube noses are mixed gliders and flappers. The fulmars, shearwaters and large petrels are capable of feeding on the wing, snatching food from the surface, but they feed more frequently while swimming on the surface than the albatrosses. The small storm petrels almost walk on the water as they flit over the waves. Only the diving petrels catch their food by diving. Many of these attractive little birds are almost helpless on land and nest in burrows where they are easily killed by cats or rats.

**Migrations** Many of the shearwaters and petrels perform remarkable migrations. In the Atlantic, the great shearwater, sooty shearwater and Wilson's petrel migrate well into the Northern Hemisphere during the southern winter. The great shearwater breeds on Tristan da Cunha; it crosses the equator late in April or in May and gathers in large numbers off west Europe late in the northern summer before returning south. In the Pacific, four species of shearwater, the slender-billed, the pink-footed, the pale-footed and the sooty shearwaters, undertake similar extensive migrations. The slender-billed shearwater migration follows a figure-of-eight pattern from their nesting sites off Australia to north of the Aleutians.

Several northern hemisphere breeding species migrate south in winter. Many overwinter in the upwelling regions off the Peruvian and north and south-west African coasts where the feeding is so rich. The three smaller skua species, the pomarine, long-tailed and Arctic skuas, all overwinter at low latitudes. The ocean-going grey and red-necked phalaropes also overwinter far out to sea in high concentrations, bobbing like little corks high on the water. The phalaropes are called gale birds since they are too small and light to fly against storm winds, and so they can get blown ashore in large numbers while on migration flights. The greatest migration of all is that of the Arctic tern, which flies almost from one pole to the other.

**Swimming and diving** Sea-birds feed either at the surface or by diving. Surface-feeders—whether they feed while sitting on the surface or by skimming or snatching food while in flight—fare much better if they feed at night. Optimum times will be at first and last light on moonless or cloudy nights, as this is when the vertically migrating plankton swims right up into the surface layers.

In contrast, diving birds which hunt far more by sight, are diurnal feeders. Diving birds have less pneumatic skeletons than other species; they exhale on diving and their plumage compresses. These adaptations contribute towards making their density closer to that of water when they submerge. Cormorants are the only species in which the plumage actually becomes sodden—this is why they spend so much time drying their wings. Even very similar species of diving birds may have feeding habits which result in their exploiting quite different food resources. In Europe, the shag, *Phallacrocorax aristotelis*, and the northern cormorant, *P. carbo*, overlap in their breeding ranges. The shag feeds in midwater on sand-eels and other fishes; the cormorant feeds close to the bottom, mostly on small flat-fishes. Clearly, any bird feeding close to the bottom is restricted to inshore areas where the water is relatively shallow.

There are several plunging birds. Brown pelicans, *Pelecanus occidentalis*, catch fishes by oblique dives. Tropic birds and terns plunge from heights of 20 m, gannets and boobies from more than 30 m.

**Pirates** Several species of sea-birds are part-time pirates. In polar regions, the great skuas are pillagers and robbers of nesting sites. They harry other bird species until they void their crops full of food which the skua then swoops on and eats. In the northern hemisphere, the three smaller skua species vigorously chase fulmars and kittiwakes; the kittiwakes are the only true ocean-going gulls. In the tropics, the frigate birds mercilessly harry boobies returning to their nests after foraging trips. They rob them of food and nest-building material, and then rob each other. A twig may pass from bird to bird in great aerial dog-fights. Frigate birds otherwise are expert snatchers of food from the sea surface. They are surprisingly light with their large wing span and can glide and soar with superb grace. This conquest of the air, far from denying birds the opportunity of exploiting the ocean's resources, has opened up safe and reliable supplies. However, being the end of the food chain does render birds particularly susceptible to the accumulation of organochloride residues and heavy metals. The brown pelican populations around the coasts of the U.S.A. were almost wiped out when DDT residue levels in their bodies rose so high that the eggs which they laid were so thin-shelled that they were crushed by the brooding birds. Luckily the danger was seen in time and controls in the use of DDT have led to some of the brown pelican populations beginning a recovery.

PATON

The European oystercatcher (*Haematopus ostralegus*) illustrates the conflict between conservation and exploitation: because they feed on commercial cockle beds, large numbers are shot. Conservationists claim that these control measures are ineffective and unnecessarily disturb the habitat.

The dunlin (*Calidris alpina*) shows many features common to all water birds. The long beak is used to probe for worms and small shellfish living in intertidal mudflats. Each species has a beak of slightly different length and so exploits different food animals.

VAN DER KAM

The fulmar (*Fulmarus glacialis*) is a truly oceanic bird. Its stiff wings have a dihedral cross-section that enables it to soar and glide, using the uplift above waves. Their tube-noses are typical of their family, which includes albatrosses and shearwaters.

ZIESLER

HEATHER ANGEL

Galapagos sea lions (*Zalophus californianus wollebaeki*) are marine mammals and so suckle their young. The thick pelt and a layer of fatty blubber help to keep the animals from losing too much heat in the water.

Right: The southern fur seal (*Arctocephalus tropicalis*) was almost exterminated by sealers. Now fully protected, the South Georgia colony is growing so rapidly that it threatens other species. The population explosion possibly results from the superabundance of krill that has become available since the reduction of whale stocks.

## Dugongs and manatees

The sea-cows are dolphin-like mammals which grow up to 3 m long and 500 kg in weight. The fore limbs are flipper-like, but there are no hind limbs. The body is spindle-shaped (fusiform) with the tail expanded into a fluke-like organ. The skin is thick, sparsely covered with hair and underlain by thick blubber. The surviving species are all slow placid animals, living in tropical waters and feeding on eel grasses and turtle grasses. Their skeletons are solid and, since they have no bone marrow, it is a puzzle from where their red blood corpuscles originate. They never ever come out on land and even the young are born underwater.

The largest species was Steller's sea-cow. This huge seaweed eater weighed 4–5 tonnes and was the only cold water species. Twenty-seven years after its discovery off Alaska, it was exterminated, although a recent Russian report suggests there is just a possibility that a few may still exist. The only truly marine species of sea-cow is the dugong. These animals are abundant now only off parts of South-east Asia and North Australia. They have suffered partly because they are good eating, but also because of their meat's reputed aphrodisiac properties and the medicinal uses of the fat and oil. Manatees are riverine and estuarine species. In the sea they eat mostly eel grass. They congregate in large parties by day and scatter to feed at night.

## Sea-lions, walruses and seals

This group of mammals is almost totally marine and needs to come on to land or ice only for breeding. Their streamlined fusiform bodies, lined with a thick insulation of blubber and their flipper-like limbs, are all adaptations to an aquatic life. They are carnivores, feeding on fishes, squids or crustaceans in midwater, or on bottom-living animals. Each species has its own particular diet and pattern of feeding. The eyes have a flat cornea which allows them to focus well underwater. Out of

DOSSENBACH/NSP

water the iris closes down and the eye then acts rather like a pin-hole camera. Since some seals are perfectly able to survive when blinded, they must use clicks and whistles for underwater navigation.

As a group, the seals have relatively few enemies. In warm waters, sharks are their main danger. In the colder high latitude waters where most species live, killer whales and leopard seals are their chief predators in the water but the most persistent and vicious killer of seals is man. Vast populations of several species of seals have been reduced to the verge of extinction by greedy exploitation.

**Eared seals** This family include sea-lions and fur seals. They have small external ears. When swimming they use their fore flippers for propulsion and the hind flippers for steering. On land the fore flippers turn outwards and the hind flippers turn forward, so the seals can run at a lumbering gallop. There is a very marked sexual dimorphism, the males being much bigger and heavier than the females. Bull Steller's sea-lions weigh nearly a tonne whereas the cows are a quarter that weight. This species, which is the largest of all sea-lions, eats squids. During summer they migrate high into the Arctic and move south again as pack ice begins to reform.

The greatest concentrations of fur seals in the world occur on the Pribilof Islands, which are north of the Aleutian Islands between Alaska and the Pacific coast of Russia. After their discovery in 1786, sealers began massacring the seals for their pelts. The population of two million was reduced and in 1806 sealing temporarily halted. The seals recovered, but again in 1834 sealing stopped. When it began again the fur seals again declined in numbers and killing the seals in the water was banned. Now the population is back at the two million mark, despite the annual cull of 60–70,000 male fur seals. This shows how a carefully regulated cull can still allow a population of animals to flourish and be maintained as a valuable resource. It has been calculated that the fur seals around Alaska alone eat about 700,000 tonnes of fish each year.

The Antarctic or Kerguelen fur seal became almost extinct. This is a krill-eater that inhabits all the Antarctic Islands lying near the Antarctic Convergence. On South Georgia, sealers almost exterminated the fur seals, but a small colony survived. Now this colony is producing 60,000 pups a year and may soon outnumber the northern fur seals. A possible reason for this is the abundance of their staple food, krill, since the whaling fleets have reduced the whale population so drastically. However, no similar population outburst of these fur seals has occurred on the other Antarctic Islands, so the reason may not be quite so simple. Other fur seal species occur around South America, South Africa, South Australia, Tasmania and New Zealand, all in the southern hemisphere.

**Walruses** The walrus is closely related to the eared seals, but the one species is restricted to the Arctic, where it is circumpolar. The populations in the Atlantic and Pacific differ enough to be considered different races. They congregate in massive herds on ice-floes and beaches from which they dive to depths of about 80 m. They feed on the bottom, grubbing out clams with their long tusks—which are long canine teeth. Walruses played an important role in the economy of the Eskimos, providing them with oil, meat, bone and skin. The tusks, being pure dentine, were the main source of ivory in Europe in the 16th century when the 'seahorses' were first hunted off Spitzbergen.

**True seals** The eighteen species of true seals probably represent a quite separate evolutionary branch to the other seals. The fore flippers are small and the hind flippers cannot be turned forward, so on land they are much less mobile. Swimming underwater is achieved by sculling with the hind flippers; the fore flippers are held along their flanks. These seals haul out to mate and pup. The pups are usually born with a thick woolly coat that is moulted before they take to the sea.

Southern elephant seal pups are suckled for only three weeks, on rich, fatty milk. The pup may put on weight by as much as 10 kg a day, while the cow loses a total of 300 kg during suckling. Once weaned, the pups are deserted and they stay on the breeding grounds for another two weeks until they have completed their moult. The pups take to the sea where they start to feed on crustaceans before switching to the adult diet of squids. There is a northern elephant seal that lives on the islands off California and Mexico. It is a much more placid animal than the noisy aggressive southern elephant seal of southern South America and the Antarctic.

In Antarctica, the seals that live amongst the pack ice have escaped exploitation by man. The crabeater seals feed mostly on krill, so occur to the south of the Antarctic Convergence, usually in the vicinity of the pack ice. They are circumpolar and extremely abundant; the world population is possibly over 5 million. Adults are often marked with small scars around their head and shoulders; the results of inter-seal rivalry during breeding. Many also have half metre long parallel scars, the souvenirs of encounters with their main enemies, killer whales. Weddell seals which dwell within sight of the Antarctic coast are too deep within the pack ice to suffer much from the predations of killer whales. Instead, they usually die when their teeth wear out. During winter they live below the ice, biting and sawing through it with their canine teeth to keep their blow holes open, feeding mostly on invertebrates from off the sea bed.

Ross seals are very poorly known. They live amongst the thickest most inaccessible pack ice. They have huge eyes but rather tiny mouths with needle-like teeth. Their diet consists mainly of squids and some krill. It is impossible to estimate the abundance of these animals as they are solitary and live in such an inaccessible habitat.

WALKER/NSP

The leopard seal is also solitary, usually occurring in the outer fringes of the pack ice but sometimes turning up on the beaches of South Australia and New Zealand. On land, it moves grotesquely by contorting itself like a looper caterpillar instead of by using its limbs. Leopard seals are loathed partly because of their appearance, partly because they are accused of occasional unprovoked attacks on man, and partly because of their carnivorous diet of penguins, fishes and other seals. They can shake a penguin carcase completely out of its skin.

Only three species of seal occur in warm tropical seas. All are monk seals and all are under threat of extinction. Rarest of all is the Caribbean monk seal which was last seen off Jamaica in 1949. The Laysan or Hawaiian monk seal breeds on five atolls of the Leeward Chain of islands, north-west of Hawaii. They probably number less than 2,000 animals. The Mediterranean monk seal is only a

A young elephant seal, still moulting its juvenile coat, has been deserted by its parents. Hunger is beginning to force it to leave the safety of the land to search for its food in the shallow seas around the breeding ground.

little more abundant.

In the Arctic there are several species of true seals. Species that can penetrate deep into pack ice are mostly circumpolar in their distribution. The bearded seal is a bottom-feeder that uses its long moustaches to find its food in the mud of the seabed. The ring seal feeds either in midwater on amphipods or close to the bottom on crustaceans and small fishes in shallow water. Those seals which do not reach such high latitudes, either have different subspecies in the Pacific and Atlantic, as in the common (harbour) seal, or different species occur in similar habitats in the two oceans.

**Diving** A seal's ability to dive is essential to its survival. A grey seal was caught on a line at a depth of 130 m and a harp seal was caught in a net at 270 m. The deepest divers are the Weddell seals, which regularly dive to over 450 m and can remain submerged for up to 43 minutes. When a seal dives it exhales. It relies on the oxygen it has stored in its tissues, which are rich in myoglobin. This is a respiratory pigment that acts, like haemoglobin in the blood, as a reservoir for oxygen. Seal's flesh is very dark in colour because of the abundance of the myoglobin. Oxygenated blood is also stored in special blood vessels collectively known as retia; the blood volume is very much greater than in other mammals. As soon as a dive starts, the heart beat slows dramatically. Blood supply to most of the body is cut off, only the brain is supplied with oxygenated blood stored in the retia. Seals avoid the 'bends'—the decompression sickness which affects human divers if they re-surface too quickly—partly because they exhale on diving, and also because the alveoli in the lungs collapse under pressure. The alveoli are the thin-walled sacs across which gas exchange occurs between the blood and the air in the lungs. The air from the alveoli is forced into the trachea, which is held open by supporting rings of cartilage.

**Heat regulation** Seals are mammals and therefore warm-blooded animals. They keep their blood temperature at about 38°C. Some species, like the fur seals, have a thick pelage that traps within it a thin insulating layer of air when the seal dives. Typically a fur seal has a thick long guard hair emerging from each hair canal together with 19 finer underhairs. It is this soft warm underfur that makes sealskin coats so desirable. Other species have little or no hair and rely on their blubber for insulation; the walrus has a hide 5 cm thick with 15 cm of blubber beneath. Blubber, which is a source of oil, is another seal commodity avidly sought by the hunters.

If a seal is highly active, it will get hot. Unless it can lose this heat, it could suffer from overheating. This problem is particularly acute when they haul out on to land to breed. When northern fur seals are on their breeding ground, they can be seen fanning with their enormous hind flippers. Underwater, the flippers are not normally kept at body temperature, because they are not insulated and so too much heat would be lost. The blood supply going to the flippers runs close alongside the returning venous blood. The returning cool venous blood is warmed by the arterial blood which, as a result, is cool by the time it enters the flipper. This counterflow system normally helps to conserve heat, but the flipper can be flushed with blood so that heat is lost and overheating is prevented.

## Whales

The whales are the most completely marine mammals. They have highly streamlined bodies insulated with layers of blubber. Both the horizontal tail flukes and fore limbs, which are fin-like and are used like hydroplanes, possess a counterflow blood supply for heat regulation. As in the seals, the body tissues are rich in myoglobin, and oxygen is stored both in the tissues and in great reservoirs of blood in the retia. Diving is accompanied by much the same slowing of the heart and cessation of the blood supply to non-essential organs as in seals. One marked difference is that whales inhale prior to sounding whereas seals exhale.

Whales are separated into two main groups, the most ancient of which is the whalebone whales. Their earliest fossils come from deposits found in the Cretaceous era about 100 million years ago. These early forms were relatively small, ranging in size from 3–17 m, whereas modern whalebone whales are 10–33 m long. They probably evolved from relatives of an extinct group of fossil whales, the Archaeocetes. One of these Archaeocetes, *Basilosaurus*, was very snake-like in build and must have been the nearest evolution has ever come to a real sea-serpent. The other group of modern whales are the toothed whales, which include the sperm whale and dolphins.

**Whalebone whales** The whalebone whales are typified by the curtains of baleen plates that line the sides of the mouth. Their food consists of either planktonic crustaceans or fishes and is sieved out as the whale takes a great gulp of water and squirts it out through the baleen plates. Each gulp by a blue whale filters 5–6 m³ of water. In amongst the dense swarms of krill they can soon fill their stomachs with nearly half a tonne of food. The most extraordinary of all the whalebone whales is probably the bowhead or Greenland right whale. Almost a third of a bowhead's body is made up of its mouth. Within the mouth the 300 pairs of baleen plates are up to 12 m long and the animal's skull is distorted to accommodate these massive plates. Bowheads probably continually swim along with their mouths open, automatically sieving plankton out of the water. The food is periodically licked off the baleen plates with the massive tongue. The Biscayan right whale was probably the first whale species to be hunted by man, the Basques, in the open sea. A distinct population lives in the southern hemisphere, and is called the southern right whale. At sea, right whales can be immediately distinguished by their V-shaped blows; the twin plumes are 3–4 m high.

One other species can produce a double blow and that is the grey whale which occurs only in the north Pacific. During the summer these whales feed high in the Arctic and in autumn migrate south, one population down the coast of Korea, the other to Southern California. They migrate close into the shore and are a great tourist attraction as they migrate down the west coast of the United States. They move south to breed in the warm water of the shallow bays and lagoons of Southern California. In these inshore lagoons they were very vulnerable to whalers and were close to extinction when they were given complete protection in 1938. Migrations are a common feature of most whalebone whales. During the summer, they exploit the rich feeding at high latitudes, moving into warm waters when food becomes short. In this way they do not have to expend so much energy in maintaining their body temperature. Calving is usually in warm water since the newborn calves are not well insulated with blubber.

This pattern of migration has been well studied in the humpback whales which migrate northwards from Antarctica close inshore. Separate populations move up each side of the South American, South African, Australian and New Zealand coast-lines. The pattern of movement up the New Zealand coast shows that the migration north is led by females accompanied by maturing calves that are being weaned. Immature males and females then move through, followed by mature males and resting females. Last to move north are the females at a late stage of pregnancy. Throughout their stay in tropical water it appears that the whales starve, living off the reserves laid down during the summer glut of food. First to return south are newly pregnant females, presumably sensing the need to start feeding quickly. The resting females and the males are the next to return south, finally followed by the females with newly born calves; the calves having had time to thicken up their blubber prior to entering the cold Antarctic waters.

Blue whales had an alternative name of sulphur-bottoms because of the film of diatoms that grows

KOOYMAN

A grey whale cow blows as she surfaces, accompanied by her calf. These huge mammals perform spectacular annual migrations from their feeding grounds at high latitudes in the North Pacific down the west coast of America to breed in shallow lagoons off the coast of Mexico.

on their bellies while at low latitudes. Whale marks recovered in 1952 from seven whales tagged in 1935 and 1938 showed the faithfulness with which the blue whales return to their home waters; the whales carrying the marks were caught in precisely the same areas as they had been tagged. Blue whales are probably the largest animals that have ever lived. Males of over 30 m have been recorded. A female 27·3 m long was weighed piecemeal and found to weigh 136 tonnes. The tongue alone weighed 4 tonnes—about the weight of an elephant.

The second largest are the fin whales, which are also krill feeders. The third of these large whalebone whales is the sei whale, which has much finer baleen plates and feeds more on smaller planktonic species. In the North Atlantic and Pacific, the blue and fin whales feed on the euphausids which are the northern equivalents of *Euphausia superba*, whereas the sei whales feed more on the copepod *Calanus finmarchicus*. It has been estimated that, before 1930, when whaling began in earnest in the Antarctic, there were about 400,000 humpbacks, blues, fins and seis. The whales would have weighed about $25 \times 10^6$ tonnes or 25,000,000 tonnes and, during the summer feeding season, would have eaten three times this weight of krill and plankton—about 750,000 tonnes a day. The removal of most of this whale population must have made a massive resource of krill available to other animals. This may be the cause of the massive population outbursts seen recently in crabeater seals, southern fur seals, and the krill-eating penguins. Even the whales themselves have responded as they are growing faster, maturing earlier and cows are having fewer resting seasons without breeding.

Gestation in the blue whale takes 11 months, and suckling continues for nearly a year. The calf when it is born is 6·5–8·5 m long and weighs about 3 tonnes. Within two years the blue whale's egg, weighing less than a milligramme, grows into a weaned calf weighing 26 tonnes.

Bryde's whale and the minke whale are two other whalebone whales. Bryde's whale is very similar to the sei whale but appears to be restricted to water warmer than 15°C, where they feed on shoals of fish like pilchards and anchovies. Minkes are the smallest whalebone whales, only growing to 10 m in length. They, too, are primarily fish-eaters. It is interesting that the whales that exploit food lower down in the food chain by eating herbivorous plankton are larger than those eating fish, which are further from the primary source.

**Toothed whales** The toothed whales are all fish- or squid-eaters. Many of them hunt in organized packs, coordinating their efforts by sound signals. Toothed whales not only use their calls for mating and other social activities, but also for echo-locating their food. Sperm whales, for example, produce a series of clicks of varying frequency and in trains of different intervals. Each individual whale has its own special pattern of clicks which, like a fingerprint, are specific to each individual. It must be an advantage to whales hunting in schools to identify the echoes of their own call. It seems likely that sperm whales can identify targets up to a kilometre away and they can hear each other up to ten kilometres away. Off South Africa, spotter aircraft are used to home whale catchers on to their quarry. The whales' response to the catcher's approach is interesting. The whales begin to crash their flukes down hard on the surface when the catchers are still several kilometres away. This appears to be a warning signal, because the school immediately begins to make off. Two whales were seen to sound in water over 2,000 m deep. One stayed submerged for 53 minutes, the other for 112 minutes. Both whales were caught. One contained a freshly eaten black dogfish that was almost certainly taken very close to the bottom. Previously, the deepest record for a sperm whale dive was of an animal that became tangled in a submarine cable at a depth of nearly 1,200 m off Peru. Clearly such deep dives expend much energy and a long range echo-location system is extremely valuable. If this echo-location system is disorientated mass strandings of whole schools of toothed whales can occur.

The head of the sperm whale is very bulbous and filled with a waxy substance called spermaceti. There are various suggestions as to the function of the spermaceti; one is that it acts rather like an acoustic lens in focusing the whale's echo-locating clicks. Another is that the spermaceti at the whale's blood temperature is in a phase intermediate between being solid and molten. Liquid spermaceti is less dense and so at the end of a strenuous heat-generating dive it will help to buoy the whale back up to the surface.

The biology of the sperm whale is better known than that of other toothed whales because they have been commercially exploited, giving scientists ample material on which to base their measurements. In addition, the major source of specimens of giant squids, on which the sperm whales feed, is the stomachs of whales caught commercially. The analysis of the data from 37,000 whale captures by American whalers between 1765–1920 has provided the best information available on the distribution and migrations of any whales. Data on growth rates and breeding success and results of population studies are invaluable in the proper conservation of the sperm whale stocks, especially if world demand for whale products overturns the efforts of many people to ban all whaling.

The understanding of the biology of the smaller toothed whales could also be very important to our understanding of ecological processes in the sea. For example, dolphinarium studies show that a bottlenosed dolphin needs to eat 5–15 kg food a day. Using fairly conservative estimates of their abundances, the amount of fish eaten by the world's population of toothed whales comes out at a staggering 2 million tonnes a day. Some of these are fish of commercial importance, for example, the beluga which inhabits Arctic waters eats capelin, cod, herring, haddock and flounders plus various non-commercial species; the diet varies from locality to locality.

The most notorious toothed whale is the killer whale. One early report accounted for the remains of 13 porpoises and 14 seals in a single stomach. Killers attack seals in the Antarctic by tipping them off ice-floes and in this way are always considered to be a potential danger to man. Their normal diet consists of fishes, squids, with occasional sharks, seals, birds and other small whales. They are remarkably efficient hunters. In Twofold Bay, New South Wales, a killer pack developed a fascinating symbiosis with the local whalers. The killers drove whalebone whales into the Bay, and would alert the whalers. The whalers would harpoon and kill the whalebone whales. The corpse would sink and the killers ate the tongue away. Two days later the corpse, beginning to putrefy, floated to the surface and could be towed ashore for processing by the whalers.

Dolphins, which belong to the same group as toothed whales, are reputed to be among the most intelligent of all animals. There are remarkable stories of co-operation between dolphins and man. One dolphin, called Pelorus Jack, conducted ships through the Cook Straits between the North and South Islands of New Zealand for 32 years. A Florida housewife who got into difficulties while swimming was pushed ashore by an obliging dolphin. Dolphins help their newborn calves to the surface to enable them to breathe, and aid injured adults in much the same way. This sense of co-operation must be a product of their efficient schooling behaviour.

There are several extremely curious dolphins. The narwhal is a form of spotted dolphin that occurs north of 70°N in the high Arctic. The male has a secondary sexual character of growing a single spiral tooth 2 m long. This tusk, probably the origin of the myth of unicorns, is too fragile to be used in feeding or in defensive fighting and so its function seems to be purely one of sexual display. The skin of the narwhal is extremely

HEATHER ANGEL

Dolphins are fast, intelligent hunters which operate in schools using sound to find their prey and to communicate with one another. This bottle-nosed species clearly shows the blowhole on the top of the head and has scratch marks made during intermale fights.

rich in vitamin C, and was one of the main sources of the vitamin for the Eskimos. Another curious group of dolphins is the beaked whales. Many of the species are known only by stranded specimens. They have long jaws with only the odd tooth remaining and are probably mostly squid-eaters. Two species, Longman's beaked whale and the New Zealand beaked whale, are known only from the putrefying remains of single specimens stranded on shore. Nothing is known of their lives.

## Why study the air breathers?

All the air-breathing vertebrates are at the end, or close to the end, of food chains. Many heavy metal pollutants, persistent insecticide residues and industrial contaminants (e.g. polychlorinated biphenols) tend to accumulate up the food chain. These, then, are the animals that will give the early warning of potential ecological disasters. The effects of DDT residues caused a sharp decline in the brown pelican populations because the birds laid thin-shelled or even shell-less eggs; this resulted in more sensible use of persistent insecticides. Our inability to predict the environmental consequences of any manipulation or exploitation of any ecosystem comes from the immense complexity of extrapolating from the simple responses of individual species in controlled environments to the responses of natural multi-species communities in the wild. The science of ecology is still seeking its Newton or Einstein to pare away the confusion of detail to reveal some simple universal ecological 'Law'. Without an ability to predict, the conservationist must maintain a watching brief on the ends of the food chains for the first signs of impending danger. The more that is known about the biology of each group of animals, the less likely will be baseless scares, like the alarm caused by the population outburst of the crown-of-thorns starfish. Such scares waste valuable time and resources, and also could slow down the response of authorities to the signs of real danger threatening the whole ocean.

Rich deposits of industrial raw materials lie almost untouched beneath the ocean bed. The world energy crisis in recent years has stimulated the development of techniques for tapping submarine oilfields. Here an oil rig in the North Sea burns off excess propane gas.

# Man and the Ocean

Since a very early period in his history, possibly even from the time of his emergence as a species, man has been associated with the sea. Early human settlers tended to live close to their food sources and were quick to learn that the seashore can be a well stocked natural pantry. Kitchen middens of such coastal dwelling men are often almost entirely composed of the shells of easily obtained molluscs like limpets, oysters, mussels and cockles.

The abundance of many of these sources of food in and near estuaries doubtless led to the eventual development of skills in fishing for the comparatively easily caught estuarine fish such as plaice and flounders. Ultimately, this skill could have been extended to fishing with line or spear when simple boats had been developed.

Despite this ancient association with the seashore, real exploration of the sea for purposes of science, economics and travel did not develop until much later in human history.

## Exploring the ocean

It is, perhaps, not too much of an exaggeration to say that man's greatest innate desire (next to having a full stomach) is to know where he is. This is borne out, to some extent, by the fact that map-making of some sort has been a preoccupation of civilized man for a very long time. Limited practical experience made the maps of the world produced by the Egyptians and the Chaldeans more imaginative than informative but the cartographical knowledge of the world's first great maritime nation, the Phoenicians, must have been extensive, although no material examples of their maps have survived, if, indeed, they were ever drawn. These skilled ship-builders and sailors travelled over distances which, when seen in the proper perspective of knowledge and technology, put space travel firmly into its place. Their early voyages of discovery, trade and commerce, were surpassed, in oceanographic terms, only in the fifteenth and sixteenth centuries.

It may be said that the world's first truly scientific expedition commissioned to investigate geographical, oceanographical and biological matters as well as potential commercial affairs, was Captain Cook's voyage of 1768 in the *Endeavour*. Though Cook's voyage was followed by other expeditions, it was not until the great submarine telegraph cables were laid and periodically hauled up for repair that concrete evidence was obtained for the existence of life in the ocean depths.

WOODS HOLE

In many respects deep sea research is still in its infancy because of man's inability to go to great depths to observe for himself. Deep sea manned submarines of the Alvin class are able to descend to over 1800 m.

In 1872 H.M.S. *Challenger* was equipped with the most modern oceanographic equipment of the period and set sail for the next three years as a floating marine biological laboratory. This expedition produced an almost unbelievable amount of information about the sea as a total habitat.

From this solid and well laid foundation, the science of oceanography then developed. During the remainder of the nineteenth century, several scientific vessels sailed, usually to investigate some more or less well defined problem with increasingly complex and sophisticated equipment. Perhaps most famous of them all, particularly due to her association with Captain Scott's Antarctic expeditions, is *Discovery*. The original research vessel of that name has been followed, up to the present, by a series of similarly named ships which have been responsible for unravelling many of the mysteries of oceanic ecology—especially in the Atlantic Ocean. To any list of oceanic research vessels should, undoubtedly, be added the name *Calypso*, the vessel in which Jacques Cousteau has so successfully combined the modern skills of SCUBA exploration with more traditional oceanographic techniques.

With the development of aviation it became possible to observe distributions of planktonic organisms from a height above the ocean surface sufficient to show relative densities of these organisms over considerable areas. The development of satellites has extended this type of observation very considerably and has given oceanographers a whole new body of information about the very nature and origin of the oceans themselves. Modern technology in the scientific use of photography has also extended the reach of oceanographic investigators.

Although man has, to a great extent, conquered the problems of lifting himself off the surface and away from the gravitational pull of the earth, he has not managed the converse operation nearly so well. Deep-sea manned exploration is still very much in its early stages. In 1960, Jacques Piccard and Donald Walsh descended to a depth of 10,910 m in the Challenger Deep of the Marianas Trench. Submarine vessels capable of carrying several people to great depths, equipped with cameras, radio and television have now been developed and there are underwater laboratories or 'sea labs' which rest on the bottom while divers in heated suits explore the sea-bed. Oceanographic data on tides, currents, waves, salinity, pressure and temperature as well as such phenomena as barometric pressure and wind are now obtained by remote sensors attached to anchor buoys and transmitted by radio to collection stations.

The kind of information derived from these sources adds daily to the sum of our knowledge of the oceans and enables us to increase our degree of exploitation of this vast natural resource. Unfortunately, exploitors of the seas frequently forget that they are dealing with a natural, living ecosystem of great complexity and often delicate balance. Foolish and uninformed manipulation of this balance can destroy it permanently.

## Fishing and fisheries

Whilst primitive man may have been content to scrape a sufficient number of limpets or mussels from easily reached rocks for his immediate needs, the demands of civilization and community living dictate a more organized exploitation of potential food resources.

Fixed organisms, e.g. bivalve molluscs like oysters and mussels, share with commonly cultivated garden vegetables the useful property of being transplantable! In exactly the same way that a garden of cabbages may be planted in suitable soil, in a suitable locality and then harvested at the most productive period, oysters and mussels can also be 'planted' in suitable water, in a suitably sheltered area and harvested at the most suitable moment. Shellfish culture is an old art which may well have originated, along with so many other useful skills, with the ancient Chinese. It is to Roman cultivation of the oyster, however, that almost all modern fisheries trace their origins.

Whilst it is possible to 'farm' some marine fish by trapping them in man-made lagoons and fattening them upon food concentrates in a manner which is not unlike that employed in modern 'factory farms', such techniques are difficult and, as yet, must be regarded as experimental. The majority of fishing industries prefer to leave the 'husbandry' to nature, leaving only the culling of the crop for themselves. The majority of active fish-hunting is now done more or less indiscriminately by means of a net. At its simplest, this may be seen in the cast-net of oriental shallow-water fishermen. At its most complex, net fishing is a £ multi-million industry with ocean-going ships equipped with every modern gadget from electronic devices to indicate the presence of probably profitable shoals to machines which transform the catch into fish fingers and deep freeze them on the spot ready for the supermarket! It is the modernization and sophistication of the fishing industry which has enabled catches to be increased to such an extent that some areas of the oceans are considerably overfished with an attendant (and far more worrying in the ecological sense) imbalance in breeding populations.

One 'fishery' which, due to overexploitation, has almost resulted in the extinction of a notable part of the marine ecosystem is the whaling industry. Due to the fantastic profits which could be made until comparatively recently from the sale of practically all of the carcase of one of the great

NEVILLE COLEMAN

Shellfish are farmed in suitable habitats in many parts of the world. In this Tasmanian oyster farm the seed oysters are suspended in the plankton-rich water so that they are easily accessible for cleaning and collecting.

Right: Scientists on board *Lulu*, mother ship of the submersible *Alvin,* monitor *Alvin*'s movements far below. The picture was taken during Project Famous, a joint French-American study of the geological structures of the Mid-Atlantic Ridge.

Trawling is heavy and often dangerous work. Not the least onerous of the fishmen's task is sorting the catch after shipping and unloading the nets, in order to ensure that the fish is fresh.

PHOTO AQUATICS

WOODS HOLE

NEVILLE COLEMAN

One of the most important of marine pollutants is crude oil. Shores throughout the world have increasingly been affected by oil spillage in recent years, often with disastrous consequences to the littoral community.

whales, together with its valuable by-products, these magnificent mammals have been hounded by generations of skilful harpooners and, in more recent years, by 'factory ships' of ever-increasing complexity. A certain amount of international agreement, coupled with the fact that plastics are cheaper than whalebone and chemical fixatives cheaper than ambergris, may have produced a degree of control over this senseless slaughter sufficient to enable stocks of some species of right whale—notably the blue whale—to increase again beyond extinction point. If this does not happen and the great whales become extinct, it will have been as the result of man's cupidity, stupidity and inability to comprehend that delicate balance which maintains any natural ecosystem.

Man's other great contribution towards upsetting the marine habitat's ecological balance has resulted from his carelessness as a technologist and the high-handed manner in which he uses the sea as a limitless refuse dump for his unwanted materials.

## Pollution

'The sea can take it' may well, ultimately, be inscribed upon the wreath which will be cast on the surface of the lifeless, stinking, radioactive, oil-covered soup which will lap the world's shores.

There can be no doubt that the cubic capacity of the oceans is enormous and this has led to the supposition that any waste material put into the sea will soon be at infinite dilution and, therefore, harmless. It is also beyond doubt, however, that the oceans are finite and bounded by inelastic margins. Sea-water is a moderately concentrated solution of various salts because of the constant evaporation of water from its surface. Those salts were, however, introduced into the sea in dilute solution in rivers and other surface run off from the land. In exactly the same way, industrial waste continuously fed into the sea at low concentration must ultimately be concentrated.

This is true also for radioactive waste materials. In this case the effect upon the marine ecology in the long term is likely to be worse since, whereas some of the non-radioactive substances may become effectually removed from circulation by chemical reaction producing insoluble, and therefore relatively harmless, precipitates, radioactivity is not so simply isolated. Many of the radioactive chemicals poured into the sea at the present time will remain active for tens of thousands of years.

Even if the rate of concentration of many of these waste chemicals by evaporation is low, they can be built up to toxic levels by other means. Marine algae, in common with all green plants, absorb dissolved inorganic substances and use them as raw materials for making food. Those substances may have entered the sea in the effluent from an atomic energy establishment and thus concentrates of radioactive organic materials may be built up. If the algae happen to be in the phytoplankton, those radioactive substances will, inevitably, find their way through the normal, complex food chains of the sea.

Marine animals, by the nature of the pumping, filtering systems by which many of them obtain both food and oxygen, pass very large volumes of sea-water continuously over organs and tissue systems, e.g. tentacles, gills, perforated pharynx-walls, which have evolved expressly for the purpose of facilitating ionic interchange. By this means, therefore, high levels of alien materials can build up in the tissues. The salts of heavy metals, often found in the effluent of industrial processes, are especially troublesome in this way. Such accumulation can be remarkably quick—oyster beds exposed to a very low concentration of copper salts have accumulated sufficient copper in little more than one growing season to render them useless.

It is well known that insecticide treatment for pests on land is, nowadays, the object of much debate and dissent since it is known that many of the insecticidal chemicals can be built up into the tissues of plants and animals. This can also happen in the sea. The presence of 'escape' insecticide can cause twofold damage. Whole areas of planktonic organisms may be wiped out by direct toxic effects and, in much the same way as the heavy metal ions, insecticidal chemicals or their decomposition products can find their way extensively into the food chain.

Because of its dramatic and unlovely obvious effects, oil is probably the best known marine pollutant. Annually, television screens recording the progress of some new oil slick, present the sorry spectacle of sea-birds, bedraggled and slowly and painfully dying. This arouses public outcry, and rightly so; it is, however, only one small part of the devastation wrought by oil spillage.

Whilst it is true that many organisms may be wiped out directly by the presence of an oil film, e.g. the fixed invertebrates on a rocky shore, it may be that human intervention to remove the oil may be more destructive to the habitat. Careful ecological studies after major oil spillages, have shown that, left to itself, oil covering a shore can eventually support populations of some benthic organisms. Being essentially an organic substance, the oil is also susceptible to bacterial decay so that, after long periods, even the most tar-like deposits are eroded.

Human methods of dealing with oil spills are basically of three sorts—the sinking of floating slicks by the application to them of stone dust or some similarly dense material, burning off deposited littoral oil by means of high temperature flame guns and the removal of deposited oil by the application of detergents. Each technique can cause ecological disaster! An oil slick weighted down with dense material can form a rather long-lasting and stifling blanket over the bottom community on which it settles. This, however, is probably less unacceptable than the ultimate fate of a shore from which the oil from such a slick might be removed if it beached. The application of intense heat, either to burn or to char tarry residues to a more easily manipulated form, is also very effective as an incinerator of littoral organisms! This method of oil removal has been used on sandy shores where, due to their frequently critical and restricted habitat requirements, some of the burrowing fauna may take very long periods of time to re-establish themselves after incineration.

It is the application of detergent to rocky shores which is likely to have the most long term effect. Many fixed benthic organisms have planktonic or pelagic larvae which are instrumental in selecting a setting site suitable for the further development of the organism. In some cases, e.g. the clustering barnacles *Balanus balanoides* and *Chthamalus stellatus*, the site-selecting cypris larva is attracted by a hormone-like substance given out by already established adults. This substance also has the property of being chemically bonded to the rock surface where an adult barnacle, now dead and removed, has been fixed; it is, therefore, an important part of the biology of these species which must sett within penis length of one another in order to propagate and so maintain the species. Detergent is capable of removing not only already fixed barnacles but also the chemically held attractant from rocks. Similarly, it can remove from rock surfaces the important film of bacteria and microscopic algae which assist in the setting operations of many planktonic larvae.

Some of the most far reaching effects of marine pollution are those which occur indirectly. Although the presence of radioactive substances in solution is a cause for disquiet it may be that, over long periods of time, radioactive waste which is put into the sea in an apparently harmless form will cause greater ecological harm. One method of dumping such substances is to encase them in a variety of resistant containers of steel, concrete, etc. and to sink them in the greater deeps. Such a policy is short sighted, since even the greatest deeps and abysses of the ocean are scoured by slow-moving currents which ultimately rise and mix with surface waters.

Of more immediate concern are the indirect effects which can result from the presence of industrial chemicals. The surface layers of the sea are of paramount importance since it is here that gaseous interchange occurs between the dissolved gases in the water and those of the atmosphere.

In order to carry out photosynthesis, the phytoplankton require an adequate supply of carbon dioxide and this they obtain, in the sea, as the result of a buffered reaction between carbonates and the ionic product of dissolved carbonic acid in the water and gaseous carbon dioxide in the air. This balance is a delicate one and, in common with all buffered reactions, liable to be destroyed by the presence of interfering ions.

It is pleasant to be able to conclude this gloomy catalogue with an account of a form of pollution which, whilst capable of altering local oceanographic conditions and their attendant ecology, must be regarded as more benign than the more usually quoted forms of toxic pollution. Sea-water is often used by industrial plant for cooling purposes—it is a much used coolant for coastal power stations for example. When water which has passed through a system of turbines and been heated is returned to the sea, it is capable of raising the water temperature by a significant amount, especially if it is taken from and returned to an enclosed body of water like a locked dock. The introduction into such a situation of ship-borne juveniles from a different climatic region can result in the establishment of 'foreign' zoogeographical pockets.

## Physical resources from the ocean

As would be expected of a body of water which has its origins in rivers washing through and over geological strata, sea-water contains in solution enormous quantities of a large number of elements. Some of these, such as sodium and chloride are present in relatively high concentrations whilst others, such as gold, are present only as minute traces. Some of these dissolved elements are commercially extracted, e.g. sodium chloride, magnesium and bromine, but the cost of extracting the less concentrated ions far exceeds the value of the product. In the present state of extraction technology, therefore, these substances must remain untouched despite the undoubted great wealth of rare and precious materials which might otherwise form a part of industrial economy.

The other abundant constituent of the ocean is, of course, water, itself a rare and precious commodity in many parts of the earth. Desalination is undertaken in some areas where water needs are critical but it, too, is a prohibitively expensive process.

Quite apart from the dissolved mineral resources of the sea, the sea-bed itself contains great riches. Many valuable substances precipitate out of the sea-water and come to rest on the surface of the deep ocean bed. Of these, it is likely that only the manganese nodules will form an important marketable commodity in the foreseeable future—and that possibly not for the manganese but for the smaller amounts of copper, nickel and cobalt which they also contain.

Mineral-bearing strata extend under the sea as they extend under dry land and these, whenever possible, have been worked in the past. In this way coal, tin and other substances have been mined despite the ever present danger of flooding.

Of more recent importance has been the discovery and exploitation of submarine oil and gas deposits which are seen by economists as an important factor in the present world fuel crisis. It is, in a sense, tragic that such tremendous expenditure of money and manpower has to be used to tap these fossil fuels when the sea itself is a source of unbelievable energy. The problem is that tapping that energy is extremely difficult.

The sea rises and falls tidally with considerable regularity. It would seem logical, therefore, to convert that vertical range of movement into circular movement which could drive turbines and generate electricity. Unfortunately, tidal rhythms are cyclical so that, with occasional exceptions, no two adjacent tides rise to the same height. This, and other problems too abstruse to be considered here, make the direct generation of power from tidal motion impracticable.

The only apparently feasible method of converting tidal power into electrical energy is to concentrate tidal forces so that by running through restricted channels, they can drive generators in a manner which is somewhat similar to that employed in hydroelectric systems in mountainous areas. A successful generating plant of this type is operated in Brittany.

Between the years 1962 and 1967 the estuary of the River Rance was cut off by means of a dam perforated by sluices in which were situated specially designed generators. The area to landward of the dam is flooded by tidal flow and forms a gigantic reservoir; the water, pouring through the sluices at tidal ebb, turns the generators. This system is effective because of the appreciable tidal range (average value 11·40 m) in the Rance Estuary. Despite the high cost of building and maintaining such a system, it may well be that more and more nations will, in the future, be forced to look to the sea for their sources of power.

# Bibliography


BARRATT J & YONGE C M: *Collins Pocket Guide to the Sea-shore*, Collins, London, 1958
BENNETT I: *The Fringe of the Sea*, Rigby, Adelaide, 1966, Tri-Ocean, San Francisco, 1967
BUDKER P: *The Life of Sharks*, Weidenfeld and Nicholson, London, 1971, Columbia University Press
DAKIN W J: *Australian Sea-shores*, Angus and Robertson, London, 1953, Tri-Ocean, San Francisco, 1969
DAKIN W J: *The Great Barrier Reef*, Ure Smith, Sydney, 1963
DROOP M R & FERGUSON WOOD E J: *Advances in Microbiology of the Sea Vol. 1*, Academic Press, London & New York, 1968
DUXBURY A C: *The Earth and its Oceans*, Addison-Wesley Publishing Company, Philippines, 1971
FRIEDRICH H: *Marine Biology*, Sidgwick & Jackson, London, 1968, University of Washington Press, 1970
GILBERT K & YALDWYN J: *Australian Sea-shores in Colour*, Tuttle, Rutland Vermont, 1970
HARDY A: *Great Waters*, Collins, London, 1967, Harper Row, New York, 1967
HARDY A: *The Open Sea, I. The World of Plankton II. Fish and Fisheries*, Collins, New Naturalist, London, 1956 and 1959, Houghton Mifflin, Boston, Mass., 1964
HARRISON R J & KING J E: *Marine Mammals*, Hutchinson University Library, London, 1965
HERRING P J & CLARKE M R (Eds): *Deep Oceans*, Arthur Barker Ltd, London, 1971
JOHNSON M E & SNOOK H J: *Sea-shore animals of the Pacific Coast*, MacMillan, New York, 1968
MARSHALL N B: *Explorations in the Life of Fishes*, Harvard University Press, Cambridge Mass., 1971
MILLER R C: *The Sea*, Nelson, London, 1966, Chanticleer, New York
MORTON J & MILLER M: *The New Zealand Sea-shore*, Collins, London, 1968
RAYMONT J E G: *Plankton & Producitivity in the Oceans*, Pergamon Press, London, 1963
REIDL R (Ed): *Fauna und Flora der Adria*, Paul Parey, Hamburg, 1963
RUSSELL F S & YONGE C M: *The Seas*, Frederick Warne, London, 1975
SLIJPER E J: *Whales*, Hutchinson, London, 1962, Hillary, New York, 1971
WILSON D P: *They live in the sea*, Collins, London, 1947
YONGE C M: *The Sea-shore*, Collins New Naturalist, London, 1949, Atheneum, New York, 1963


# Glossary

*abyssal* Of the deepest layers of the ocean.
*aerobic* Thriving only in the presence of oxygen.
*anaerobic* Capable of living in the absence of free oxygen.
*anastomosing* Communicating by cross connections forming a network.
*anion* An *ion* which moves to the positive pole in electrolysis.
*barbel* Beard-like appendage of the mouth in certain fish.
*bathypelagic* Of the bottom layers of the open ocean.
*benthic* Of the ocean bed.
*bilateral symmetry* A body plan wherein a dorsal and ventral surface can be distinguished.
*bioluminescence* Light produced by living organisms.
*biomass* The entire assemblage of living organisms in an area.
*byssus* Bundle of fibrous threads by which some molluscs attach themselves to the *substrate*.
*cation* An *ion* which moves to the negative pole in electrolysis.
*cheliped* Arthropod limb bearing a prehensile claw.
*cilia* Hair-like organelles occurring in large numbers and used for propulsion or creating currents.
*colloidal* Glue-like.
*colonial* Made up of interdependent and often interconnecting individuals, e.g. coelenterates.
*commensal* An organism living with one or more other organisms for the purpose of sharing food.
*demersal* On or near the ocean bed.
*epipelagic* Of the upper layers of the ocean.
*euphotic zone* The upper layer of the *photic zone*.
*eutrophication* Over-enrichment with nutrients, either naturally or as a result of pollution, resulting in a too abundant growth of plants and animals.
*exoskeleton* Skeletal elements forming the surface of an animal.
*flagella* Motor organelles resembling large *cilia* but occurring singly or in small groups.
*fusiform* Spindle-shaped.
*heterotrophic* Obtaining nourishment from organic substances.
*holdfast* Attachment organ of seaweeds.
*holozoic* Obtaining nourishment from other organisms.
*ion* An atom or group of atoms carrying an electric charge and forming one of the elements of an electrolyte.
*lipid* Fat or fat-like substance.
*littoral* Area of shore between high and low tide marks.
*lophophore* Horseshoe-shaped crown of ciliated tentacles.
*mantle* Outer soft coat of molluscs and brachiopods.
*mesopelagic* Of the middle depths of the ocean.
*nauplius* A planktonic larval stage of many crustaceans.
*nekton* Actively swimming organisms.
*nematocyst* Characteristic stinging cell of coelenterates.
*neustonic* Floating on the surface of the ocean.
*operculum* Horny plate closing the entrance to a gastropod shell.
*parapodia* Paired paddle-like appendages of polychaete worms.
*pelagic* Of the open ocean; free-swimming.
*photic zone* Surface waters of the ocean penetrated by sunlight.
*photophore* Light organ.
*phytoplankton* Plant constituent of the *plankton*.
*plankton* Drifting organisms.
*radial symmetry* A body plan wherein parts are arranged around a central point so that any cut through the centre yields two identical halves.
*radula* Rasping tongue of gastropod molluscs.
*sett* Settling of barnacle larvae on *substrate* prior to metamorphosis into the adult.
*spicule* Needle-like body, generally calcareous or silicious, found in the tissues of, or as a skeleton of, invertebrates, particularly sponges.
*sublittoral* Below low water mark.
*substrate* Solid matter forming a basis for attachment; substance on which an enzyme acts.
*supralittoral* Above high water mark.
*symbiosis* Close association between two or more organisms.
*test* Hard outer covering of an organism, e.g. sea urchin shell.
*thermocline* Boundary layer between water bodies of different temperatures.
*upwelling* A rising to the surface of cold nutrient-rich water from the ocean depths, usually in coastal regions.
*zooplankton* Animal constituent of the *plankton*.

# Index

*Numbers in italics refer to illustrations.*
*† indicates extinct*

## T



## U



## V



## W



## X



## Z